AF368914

Characterization and Processing of Complex Materials

Characterization and Processing of Complex Materials

Editor

Akira Otsuki

MDPI • Basel • Beijing • Wuhan • Barcelona • Belgrade • Manchester • Tokyo • Cluj • Tianjin

Editor
Akira Otsuki
Universidad Adolfo Ibáñez
Santiago
Chile

Editorial Office
MDPI
St. Alban-Anlage 66
4052 Basel, Switzerland

This is a reprint of articles from the Special Issue published online in the open access journal *Materials* (ISSN 1996-1944) (available at: https://www.mdpi.com/journal/materials/special_issues/ Characterization_Processing_Complex_Material).

For citation purposes, cite each article independently as indicated on the article page online and as indicated below:

LastName, A.A.; LastName, B.B.; LastName, C.C. Article Title. *Journal Name* **Year**, *Volume Number*, Page Range.

ISBN 978-3-0365-8031-9 (Hbk)
ISBN 978-3-0365-8030-2 (PDF)

© 2023 by the authors. Articles in this book are Open Access and distributed under the Creative Commons Attribution (CC BY) license, which allows users to download, copy and build upon published articles, as long as the author and publisher are properly credited, which ensures maximum dissemination and a wider impact of our publications.

The book as a whole is distributed by MDPI under the terms and conditions of the Creative Commons license CC BY-NC-ND.

Contents

About the Editor

Akira Otsuki

Akira Otsuki is a Professor of Facultad de Ingeniería y Ciencias, Universidad Adolfo Ibáñez, Chile, and a Guest Professor specializing in non-destructive characterization and selective material recovery at Lulea University of Technology, Sweden. He is a member of several academic societies, including The Japan Society of Vacuum and Surface Science and the Electrokinetic Society of Japan. He has been acting as a journal editor and organizing Special Issues of journals relevant to non-destructive characterization and selective material recovery. His research is dedicated to the characterization and processing of complex materials, including colloids, waste materials, and natural ores.

Preface to "Characterization and Processing of Complex Materials"

This reprint provides a forum for scientists and engineers to share and discuss their pioneering/original findings or insightful reviews on the "Characterization and Processing of Complex Materials". Reports on successfully coupling complex industrial materials/waste/natural ores with non-destructive characterization method(s) and the unique processing developments were particularly encouraged. On the other hand, any kind of contribution under the broad framework of the understanding and production of complex materials was also highly regarded.

This reprint gathers papers on the characterization and processing of complex materials with the following keywords: non-destructive characterization; analytical chemistry and instrumentation; waste management and recycling; selective material recovery/processing; sustainability and circular economy. This includes the waste management and recycling of complex materials (e.g., food and electronic waste) through the development of a new characterization method and/or process intensification. For example, the proper characterization of complex/heterogeneous materials is still an extremely challenging task, since the majority of characterization methods analyze either the average characteristics of the whole material or a narrow area of specific interest. On the other hand, the efficient processing/recycling of complex materials/waste should be strongly supported by their characterization from the viewpoint of economic and environmental concerns.

Akira Otsuki
Editor

 *materials*

Editorial

Editorial for the Special Issue: "Characterization and Processing of Complex Materials"

Akira Otsuki [1,2,3]

1 Facultad de Ingeniería y Ciencias, Universidad Adolfo Ibáñez, Diagonal Las Torres 2640, Peñalolén, Santiago 7941169, Chile; akira.otsuki@uai.cl
2 Waste Science & Technology, Luleå University of Technology, SE 971 87 Luleå, Sweden
3 Neutron Beam Technology Team, RIKEN Center for Advanced Photonics, RIKEN, Wako 351-0198, Japan

1. Introduction

The Special Issue aimed to provide a forum for scientists and engineers to share and discuss their pioneering/original findings or insightful reviews on the "Characterization and Processing of Complex Materials". Reports on the achievement of coupling the complexity of industrial materials/wastes/natural ores with non-destructive characterization method(s) and the unique advancement of their processing were particularly encouraged. On the other hand, any kind of contribution under the broad framework of the understanding and production of complex materials was also highly regarded.

The Special Issue was intended to gather papers on the characterization and processing of complex materials with the following keywords: non-destructive characterization; analytical chemistry and instrumentation; waste management and recycling; selective material recovery/processing; sustainability and circular economy. One of them is the waste management and recycling of complex materials (e.g., foods and electronic wastes) via the development of a new characterization method and/or process intensification. For example, the proper characterization of complex/heterogeneous materials is still an extremely challenging task, since the majority of characterization methods analyze either the average characteristics of the whole material [1] or a narrow area of specific interest [2,3]. On the other hand, the efficient processing/recycling of complex materials/wastes should be strongly supported by their characterization from the viewpoint of economic and environmental concerns [2].

2. The Special Issue

The Special Issue contains 10 papers from around the globe on the topics related to the "Characterization and Processing of Complex Materials", and specifically discusses the following contents.

Wu et al. [4] and Zhou et al. [5] reported their nanobubble studies. Wu et al. [4] developed a method to control the crystalline and morphology of calcium carbonate with the aid of nanobubbles and showed the structure transition from vaterite to calcite. They explained that nanobubbles can coagulate with calcite, work as a binder between calcite and vaterite and accelerate the transformation from vaterite to calcite. They also explained their proposed mechanism by the potential energy calculated using the extended DLVO (Derjaguin–Landau–Verwey–Overbeek) theory. Zhou et al. [5] studied five kinds of nanobubbles (N_2, O_2, $Ar + 8\%H_2$, air and CO_2) in deionized water and a salt aqueous solution prepared using the hydrodynamic cavitation method. They measured the mean size and zeta potential of the nanobubbles using a light scattering system, while the pH and Eh of the nanobubble suspensions were measured as a function of time. The nanobubble stability was predicted and discussed by the total potential energies between two bubbles by the extended Derjaguin–Landau–Verwey–Overbeek (DLVO) theory.

Citation: Otsuki, A. Editorial for the Special Issue: "Characterization and Processing of Complex Materials". *Materials* **2023**, *16*, 3830. https://doi.org/10.3390/ma16103830

Received: 18 May 2023
Accepted: 18 May 2023
Published: 19 May 2023

Copyright: © 2023 by the author. Licensee MDPI, Basel, Switzerland. This article is an open access article distributed under the terms and conditions of the Creative Commons Attribution (CC BY) license (https://creativecommons.org/licenses/by/4.0/).

Fukushima et al. [6,7] reported their studies on the modeling and simulation of electrical disintegration. Fukushima et al. [6] developed a simulation method using equivalent circuits of granite to better understand the crushing process with high-voltage (HV) electrical pulses that have been previously identified as a selective liberation method. Using their simulation works, they calculated the electric field distributions, produced heat and temperature changes in granite when an electrical pulse was applied. Their results suggested that the void volume in each mineral is important in calculating the electrical properties and confirmed that considering both the electrical conductivity and dielectric constant of granite can more accurately represent the electrical properties of granite under HV electric pulse application. Fukushima et al. [7] prepared equivalent circuit models in order to represent the electric and dielectric properties of minerals and voids in a granite rock sample, considering HV electrical pulse application. Using the electric and dielectric properties and the mineral compositions measured, they prepared ten patterns of equivalent circuit models. The values of the electric circuit parameters were determined from the known electric and dielectric parameters of the minerals in granite. The average calculated data of the electric properties of granite agreed with the measured data.

Elphick et al. [8] applied their scanning electron microscopy (SEM)-based non-destructive imaging technique for the characterization of nanoparticles synthesized using X-ray radiolysis and the sol-gel method. They observed the interfacial conditions between the nanoparticles and the substrates by subtracting images taken by SEM at controlled electron acceleration voltages to allow backscattered electrons to be generated predominantly below and above the interfaces. They found the interfacial adhesion to be dependent on the solution pH used for the particle synthesis or particle suspension preparation, proving the change in the particle formation/deposition processes with pH as anticipated and in agreement with the prediction based on the DLVO theory.

Kanari et al. [9–12] reported their characterization results of chromite concentrate [9], copper anode slime [10,12] and alkali ferrates synthesized from industrial ferrous sulfate [11]. Their characterization consists of inductively coupled plasma atomic emission spectroscopy (ICP-AES), SEM-EDS and X-ray diffraction in order to extract information on the chemical composition, morphology and mineralogical composition of their products prepared under different conditions.

Zhong and Xu [13] reported high-temperature mechanical behaviors of SiO_2-based ceramic cores for the directional solidification of turbine blades. They conducted isothermal uniaxial compression tests of ceramic core samples on a Gleeble-1500D mechanical simulator with an innovative auxiliary thermal system and investigated the stress–strain results and macro- and micro-structures of SiO_2-based ceramic cores. They characterized the microstructures using SEM. Based on their experimental data, they established a nonlinear constitutive model for high-temperature compressive damage. The statistical results of the Weibull modulus showed that the stability of hot deformation increases with the increase in temperature. They confirmed the applicability of their model by comparing their results between the predictions of the nonlinear model and the experimental values.

Acknowledgments: The guest editor wishes to thank all the authors, reviewers and the editors who contributed to this Special Issue.

Conflicts of Interest: The authors declare no conflict of interest.

References

1. Otsuki, A.; Mensbruge, L.D.L.; King, A.; Serranti, S.; Fiore, L.; Bonifazi, G. Non-destructive characterization of mechanically processed waste printed circuit boards-particle liberation analysis. *Waste Manag.* **2020**, *102*, 510. [CrossRef] [PubMed]
2. Otsuki, A.; Chen, Y.; Zhao, Y. Characterisation and beneficiation of complex ores for sustainable use of mineral resources: Refractory gold ore beneficiation as an example. *Int. J. Soc. Mater. Eng. Res.* **2014**, *20*, 126–135. [CrossRef]
3. Otsuki, A.; Bryant, G. Characterization of the interactions within fine particle mixtures in highly concentrated suspensions for advanced particle processing. *Adv. Colloid Interface Sci.* **2015**, *226*, 37–43. [CrossRef] [PubMed]
4. Wu, Y.; Huang, M.; He, C.; Wang, K.; Nhung, N.T.H.; Lu, S.; Dodbiba, G.; Otsuki, A.; Fujita, T. The Influence of Air Nanobubbles on Controlling the Synthesis of Calcium Carbonate Crystals. *Materials* **2022**, *15*, 7437. [CrossRef] [PubMed]

5.   Zhou, Y.; Han, Z.; He, C.; Feng, Q.; Wang, K.; Wang, Y.; Luo, N.; Dodbiba, G.; Wei, Y.; Otsuki, A.; et al. Long-Term Stability of Different Kinds of Gas Nanobubbles in Deionized and Salt Water. *Materials* **2021**, *14*, 1808. [CrossRef] [PubMed]

6.   Fukushima, K.; Kabir, M.; Kanda, K.; Obara, N.; Fukuyama, M.; Otsuki, A. Simulation of Electrical and Thermal Properties of Granite under the Application of Electrical Pulses Using Equivalent Circuit Models. *Materials* **2022**, *15*, 1039. [CrossRef] [PubMed]

7.   Fukushima, K.; Kabir, M.; Kanda, K.; Obara, N.; Fukuyama, M.; Otsuki, A. Equivalent Circuit Models: An Effective Tool to Simulate Electric/Dielectric Properties of Ores—An Example Using Granite. *Materials* **2021**, *14*, 4549. [CrossRef] [PubMed]

8.   Elphick, K.; Yamaguchi, A.; Otsuki, A.; Hayagan, N.L.; Hirohata, A. Non-Destructive Imaging on Synthesised Nanoparticles. *Materials* **2021**, *14*, 613. [CrossRef] [PubMed]

9.   Kanari, N.; Allain, E.; Filippov, L.; Shallari, S.; Diot, F.; Patisson, F. Reactivity of Low-Grade Chromite Concentrates towards Chlorinating Atmospheres. *Materials* **2020**, *13*, 4470. [CrossRef] [PubMed]

10.   Kanari, N.; Allain, E.; Shallari, S.; Diot, F.; Diliberto, S.; Patisson, F.; Yvon, J. Thermochemical Route for Extraction and Recycling of Critical, Strategic and High-Value Elements from By-Products and End-of-Life Materials, Part II: Processing in Presence of Halogenated Atmosphere. *Materials* **2020**, *13*, 4203. [CrossRef] [PubMed]

11.   Kanari, N.; Ostrosi, E.; Diliberto, C.; Filippova, I.; Shallari, S.; Allain, E.; Diot, F.; Patisson, F.; Yvon, J. Green Process for Industrial Waste Transformation into Super-Oxidizing Materials Named Alkali Metal Ferrates (VI). *Materials* **2019**, *12*, 1977. [CrossRef] [PubMed]

12.   Kanari, N.; Allain, E.; Shallari, S.; Diot, F.; Diliberto, S.; Patisson, F.; Yvon, J. Thermochemical Route for Extraction and Recycling of Critical, Strategic and High Value Elements from By-Products and End-of-Life Materials, Part I: Treatment of a Copper By-Product in Air Atmosphere. *Materials* **2019**, *12*, 1625. [CrossRef] [PubMed]

13.   Zhong, J.; Xu, Q. High-Temperature Mechanical Behaviors of SiO_2-Based Ceramic Core for Directional Solidification of Turbine Blades. *Materials* **2020**, *13*, 4579. [CrossRef] [PubMed]

Disclaimer/Publisher's Note: The statements, opinions and data contained in all publications are solely those of the individual author(s) and contributor(s) and not of MDPI and/or the editor(s). MDPI and/or the editor(s) disclaim responsibility for any injury to people or property resulting from any ideas, methods, instructions or products referred to in the content.

Article

Long-Term Stability of Different Kinds of Gas Nanobubbles in Deionized and Salt Water

Yali Zhou [1], Zhenyao Han [1], Chunlin He [1], Qin Feng [1], Kaituo Wang [1], Youbin Wang [1], Nengneng Luo [1], Gjergj Dodbiba [2], Yuezhou Wei [1], Akira Otsuki [3,4] and Toyohisa Fujita [1,*]

[1] College of Resources, Environment and Materials, Guangxi University, Nanning 530004, China; 1815394050@st.gxu.edu.cn (Y.Z.); 1622303001@st.gxu.edu.cn (Z.H.); helink@gxu.edu.cn (C.H.); fengqin307@gxu.edu.cn (Q.F.); wangkaituo@gxu.edu.cn (K.W.); wangyoubin@gxu.edu.cn (Y.W.); nnluo@gxu.edu.cn (N.L.); yzwei@gxu.edu.cn (Y.W.)
[2] Graduate School of Engineering, The University of Tokyo, Bunkyo 113-8656, Japan; dodbiba@g.ecc.u-tokyo.ac.jp
[3] Ecole Nationale Supérieure de Géologie, Geo Ressources UMR 7359 CNRS, University of Lorraine, 2 Rue du Doyen Marcel Roubault, BP 10162, 54505 Vandoeuvre-lès-Nancy, France; akira.otsuki@univ-lorraine.fr
[4] Waste Science & Technology, Luleå University of Technology, SE 971 87 Luleå, Sweden
* Correspondence: fujitatoyohisa@gxu.edu.cn

Citation: Zhou, Y.; Han, Z.; He, C.; Feng, Q.; Wang, K.; Wang, Y.; Luo, N.; Dodbiba, G.; Wei, Y.; Otsuki, A.; et al. Long-Term Stability of Different Kinds of Gas Nanobubbles in Deionized and Salt Water. *Materials* **2021**, *14*, 1808. https://doi.org/ 10.3390/ma14071808

Academic Editor: Daniela Kovacheva

Received: 8 March 2021
Accepted: 2 April 2021
Published: 6 April 2021

Publisher's Note: MDPI stays neutral with regard to jurisdictional claims in published maps and institutional affiliations.

Copyright: © 2021 by the authors. Licensee MDPI, Basel, Switzerland. This article is an open access article distributed under the terms and conditions of the Creative Commons Attribution (CC BY) license (https:// creativecommons.org/licenses/by/ 4.0/).

Abstract: Nanobubbles have many potential applications depending on their types. The long-term stability of different gas nanobubbles is necessary to be studied considering their applications. In the present study, five kinds of nanobubbles (N_2, O_2, Ar + 8%H_2, air and CO_2) in deionized water and a salt aqueous solution were prepared by the hydrodynamic cavitation method. The mean size and zeta potential of the nanobubbles were measured by a light scattering system, while the pH and Eh of the nanobubble suspensions were measured as a function of time. The nanobubble stability was predicted and discussed by the total potential energies between two bubbles by the extended Derjaguin–Landau–Verwey–Overbeek (DLVO) theory. The nanobubbles, except CO_2, in deionized water showed a long-term stability for 60 days, while they were not stable in the 1 mM (milli mol/L) salt aqueous solution. During the 60 days, the bubble size gradually increased and decreased in deionized water. This size change was discussed by the Ostwald ripening effect coupled with the bubble interaction evaluated by the extended DLVO theory. On the other hand, CO_2 nanobubbles in deionized water were not stable and disappeared after 5 days, while the CO_2 nanobubbles in 1 mM of NaCl and $CaCl_2$ aqueous solution became stable for 2 weeks. The floating and disappearing phenomena of nanobubbles were estimated and discussed by calculating the relationship between the terminal velocity of the floating bubble and bubble size.

Keywords: extended DLVO theory; mean size; zeta potential; Ostwald ripening effect; Stokes equation

1. Introduction

Nanobubbles have some unique properties, unlike conventional milli- to micro-bubbles, such as high mass transfer [1], long-term stability [2–10], high zeta potential, high surface to volume ratio, and generating free radicals when collapsing [11,12]. Nanobubbles can be divided into surface nanobubbles absorbed on solid surfaces and bulk nanobubbles dispersed in aqueous solutions, experiencing Brownian motion. Bulk nanobubbles have diameters of less than 1 micrometer [13]. Because of their unique physico-chemical properties, nanobubbles can be used in various application fields, e.g., improvement of plant growth and productivity [14], membranes cleaning [15–17], waste-water treatment [1,18–22], visualization improvement as the ultrasound contrast agent [23], froth flotation [24,25], improvements of methane production in the anaerobic digestion [26,27], applications in food processing [28,29] and reactions with concrete using CO_2 nanobubbles [30,31].

Different kinds of nanobubbles have different application potentials. H_2 nanobubble gasoline blends can improve combustion performance, compared with conventional gasoline [32]. A N_2 nanobubble water addition can enhance the hydrolysis of waste activated sludge and improve methane production in the process of anaerobic digestion [26]. O_2 nanobubbles produce the methane in anaerobic digestion of cellulose [33] and CO_2 bulk nanobubbles can be used in food processing [29].

Our group also reported free radical degradation in water using different kinds of nanobubbles, i.e., H_2 in Ar, O_2, N_2, CO_2 and a mixture of H_2 in Ar and CO_2. The hydroxyl radical was scavenged, and the superoxide anion was diminished by mixing the carbon dioxide nanobubbles after hydrogen nanobubbles existence in the water [11]. The antioxidant effect of H_2 nanobubbles in water was found to suppress tumor cell growth [12]. In those applications, it is important to investigate the stability and quality of nanobubbles in water, depending on time for further utilization.

Since there are so many potential applications, it is important to study fundamental characteristics and properties of bulk nanobubbles in complex solutions. The low concentration of 1 mM (milli mol/L) of NaCl could stabilize O_2 micro- and nano-bubbles for at least one week [7] because there was a shielding effect in low concentrations of NaCl. On the other hand, the higher concentration of NaCl decreased the nanobubble concentration more quickly [7]. Nirmalkar et al. (2018) performed the calculation of the interaction energies between air nanobubbles in 0–20 mM NaCl solutions by using the Derjaguin–Landau–Verwey–Overbeek (DLVO) theory. According to their calculation, the interaction potential energy was positive above pH 4 in deionized water and it decreased with an increasing salt concentration. Beyond a certain critical concentration of NaCl (between 10 to 20 mM), the system becomes unstable [8]. Meegoda et al. (2019) calculated the electrical double layer potential between O_2 nanobubbles in 0–0.1 M NaCl solutions and reported the stability of nanobubbles in 1 mM NaCl concentration [9]. Hewage et al. (2021) also reported the stability of air nanobubbles in 1 mM of different ion valence electrolyte solutions (i.e., NaCl, $CaCl_2$, $FeCl_3$) was confirmed over one week [10]. However, these research studies are only partial. In order to have a better and fuller understanding of nanobubble characteristics and have a reasonable interpretation of nanobubble behaviors in different solution environments, it is important to have a comprehensive study.

In the current study, five different kinds of nanobubbles of N_2, O_2, 8% H_2 in Ar, CO_2 and air were compared with their existence period, average size, and zeta potential, as well as their suspension pH and Eh as a function of time to be considered in several applications. The property changes of nanobubbles in the presence of salt as a function of time were observed and discussed. The stability of nanobubbles via our experimental studies was interpreted and discussed by using the extended DLVO theory.

2. Materials and Methods

2.1. Materials

Deionized water with a resistivity of 18.2 MΩcm prepared by the Classic Water Purification System from Hitech instruments CO., Ltd. (Shanghai, China) was used for all the experiments. Different gases (N_2, O_2, Ar + 8%H_2 (8% H_2 and 92% Ar mixed gas), air and CO_2) were used to prepare nanobubble water. Sodium chloride (NaCl), calcium chloride ($CaCl_2$), aluminum chloride hexahydrate ($AlCl_3.6H_2O$) were used to prepare 1 mM salt solutions and nanobubble water in order to study the nanobubbles' stability in the case of salt addition, considering the Schultze Hardy rule, i.e., $CCC \propto \frac{1}{z^6}$ where CCC is the critical coagulation concentration, and z is the ionic valence [34]. Dilute HCl or NaOH was used to adjust the pH of the nanobubble water. A 10 L glass container was used to store the nanobubble water temporarily and 50 mL plastic centrifuge tubes were used to store the nanobubble water after they were rinsed well with deionized water.

2.2. Generation of Different Kinds of Nanobubbles

Nanobubble water was produced by a ultra-micro bubble generator (XZCP-K-1.1) (100 nm to 10 μm) provided by Xiazhichun Co. Ltd. (Kunming, China). When the generator was turned on, gas from a gas cylinder was pumped into the machine in a negative pressure, and inlet and outlet pipes were immersed in deionized water/electrolyte solutions in the 10 L glass container where there was a circulation between the inlet and outlet, as shown in Figure 1A. The machine mixed the gas and liquid, and then high-density, uniform and "milky" nanobubble water (Figure 1B) was produced through a nozzle by the hydrodynamic cavitation method. The cloudy and milky nanobubble water gradually became clear through the process of the microbubbles rising and collapsing at the air–water interface (Figure 1C). This process took 2 to 3 min (Figure 1D). The machine worked for 15 min to generate high number density nanobubble water of more than 10^8 bubbles/mL (obtained from NanoSight, NS300, Malvern (Worcestershire, UK) outsourcing).

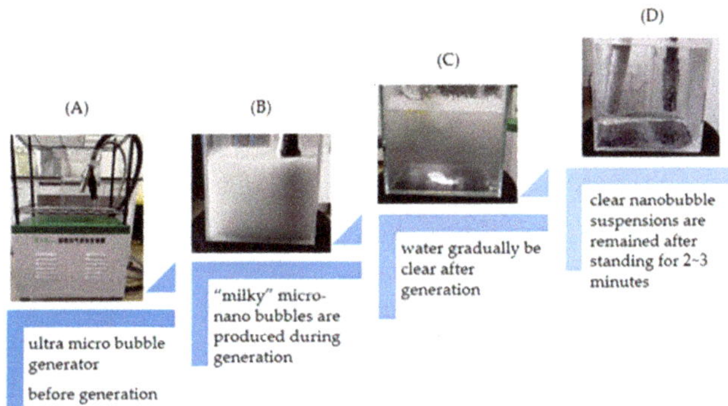

Figure 1. Procedure of nanobubble generation. (**A**) Before generation; (**B**) During generation of micro-nano bubbles; (**C**) Stop the generation of bubbles; (**D**) After standing for 2 to 3 min.

After the above-mentioned preparation steps, the nanobubble water had a temperature of about 313 K. It was then left to cool down to room temperature. Next, from the prepared nanobubble water, the large sizes of the bubbles, from 1 to 10 μm, were removed by centrifugal treatment in the following manner. The nanobubble water was stored in a 50 mL centrifuge tube; then, centrifugal treatment was performed in 6000 rpm, i.e., 31.4 m/s peripheral velocity, for 6 min to remove possible impurities and large bubbles. After centrifugation, the size distribution, zeta potential, Eh and pH were measured from one centrifuge tube to record the first day's data while other sealed centrifuge tubes with nanobubble water were kept at room temperature (298 K) for continuous measurements.

2.3. Characterization of Bulk Nanobubble Suspensions

The nanobubble size was measured by the dynamic light scattering method (DLS, NanoBrook Omni, Brookhaven Instruments, Holtsville, NY, USA). The machine's measurement range is between 1 nm and 10 μm. Since nanobubbles experience Brownian motion, the scattered light fluctuates as a function of time due to the particles' random movements, and the diffusion coefficient can be obtained from a computer digital correlator. The particle hydrodynamic diameter can be calculated by the Stokes–Einstein equation [35] as described below:

$$D_T = \frac{KT}{3\pi\eta d_h} \tag{1}$$

where D_T, k, T, η, d_h is the diffusion coefficient, Boltzmann constant, liquid absolute temperature, viscosity, and hydrodynamic diameter, respectively.

One measurement duration time was 3 min, and the instrument performed the size measurement 3 times. After size measurements, the zeta potential was measured by the same instrument that performed the 3 measurements, consisting of 20 runs/measurement. Every experiment was performed in duplicate and then the average size and measurement error were calculated and plotted.

The zeta potential value was obtained through the micro-electrophoresis method by the phase analysis light scattering method (NanoBrook Omni, Brookhaven Instruments). The zeta potential of the bubbles in the aqueous solution with salt was calculated by using the Smoluckowski model $f(\kappa a) = 1$ in Equation (2) for $a \gg 1/\kappa$ ($\kappa a \gg 1$); however, the zeta potential of the bubbles in deionized water was calculated by using the Hückel model $f(\kappa a) = 2/3$ in Equation (2) individually for $a \ll 1/\kappa$ ($\kappa a \ll 1$) [34] because of the small bubble size. The electrophoretic mobility μ in the Henry equation is shown as follows:

$$\mu = \frac{\varepsilon_r \varepsilon_0 \zeta}{\eta} f(\kappa a) \tag{2}$$

where ε_r, ε_0 are relative permittivity and permittivity of free space, respectively, ζ is the zeta potential, η is fluid viscosity, κ is the Debye length, a is bubble radius and $f(\kappa a)$ is the Henry function [36].

The viscosity, refractive index and relative dielectric constant of water at 293 K are 1.002×10^{-2} poise, 1.332, and 80.2, respectively [37]. Nanobubble water pH and Eh values were measured by pH meter (Inlab Expert Pro, Mettler Toledo, Greifensee, Switzerland) and Eh meter (by oxidation-reduction potential meter, Inlab Redox, Mettler Toledo), respectively.

3. Results and Discussion

3.1. Zeta Potential of Different Kinds of Nanobubbles

The zeta potentials of prepared different kinds of nanobubbles (N_2, O_2, Ar + 8%H_2, air and CO_2) depending on pH were measured and are plotted in Figure 2. The pH was adjusted from the natural pH by adding NaOH or HCl to achieve the desired pH. The natural pH of nanobubbles in deionized water was between a pH of 6 to 7, except for CO_2 nanobubble suspensions. The zeta potential of N_2, O_2, Ar + 8%H_2 nanobubbles were negative at about -15 to -30 mV at a natural pH of 6 to 7, while the zeta potential of CO_2 nanobubbles in deionized water was positive at about $+10$ mV at a natural pH of 4 to 4.5. The solubility of gas, dielectric constant of gas [37] and the isoelectric point (IEP) of nanobubbles determined in this study are listed in Table 1. Since the solubility of CO_2 is high (7.07×10^{-4}, Table 1), the HCO_3^- adsorption on the CO_2 bubble surface can cause a positive zeta potential in the deionized water due to the concentration of counter ions. The CO_2 solubility phenomena can be explained in the following Equations (3)–(8) [38] assuming that CO_2 gas is dissolved in water.

$$CO_2 + H_2O \rightleftharpoons H_2CO_3 \tag{3}$$

$$\frac{H_2CO_3}{CO_2} = 1.7 \times 10^{-3} \text{ at 298 K} \tag{4}$$

$$H_2CO_3 \rightleftharpoons H^+ + HCO_3^- \tag{5}$$

$$K_{a_1} = \frac{[H^+][HCO_3^-]}{[CO_2] + [H_2CO_3]}, \ pK_{a_1} = 6.35 \tag{6}$$

$$HCO_3^- \rightleftharpoons H^+ + CO_3^{2-} \tag{7}$$

$$K_{a_2} = \frac{[H^+]\left[CO_3^{2-}\right]}{[HCO_3^-]}, \ pK_{a_2} = 10.33 \tag{8}$$

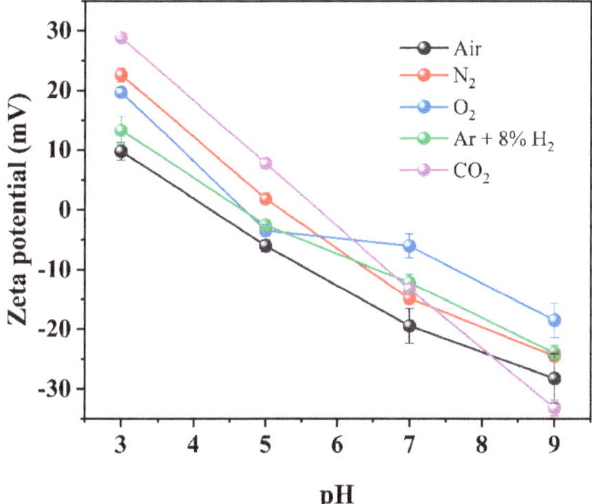

Figure 2. Zeta potential of N_2, O_2, $Ar + 8\%H_2$ and CO_2 nanobubbles in deionized water as a function of pH.

Table 1. Solubility of gas, dielectric constant [37] and isoelectric point of nanobubbles determined from the results shown in Figure 1.

Gas	Solubility of Gas in 298.15 K mol gas/mol H_2O	Dielectric Constant of Gas (Average)	Isoelectric Point (pH)
H_2	1.455×10^{-5}	1.0002532	
O_2	2.748×10^{-5}	1.0004941	4.7
$8\%H_2 + Ar$		1.0005247	4.6
Air		1.0005359	4.2
Ar	2.748×10^{-5}	1.0005360	
N_2	1.274×10^{-5}	1.0005474	5.2
CO_2	7.07×10^{-4}	1.0009217	5.7

The natural pH of CO_2 nanobubbles in the deionized water was around 4.2. The IEPs of nanobubbles in the deionized water containing air, $Ar + 8\%H_2$, O_2, N_2, and CO_2 had a pH of 4.2, 4.6, 4.7, 5.2 and 5.7, respectively (Table 1). They were determined from the two pH values of zeta potential, assuming a linear relationship. Among all the gases studied, the IEP of CO_2 was the highest (i.e., 5.7). This can be explained by the dissolution of the bubble surface, as discussed above, and because the IEP increased with the dielectric constant of gas increase, as listed in Table 1.

3.2. Time Effect of Mean Size and Zeta Potential of Different Gas Nanobubbles and pH and Eh of Nanobubble Suspensions

The effect of time after the preparation of the nanobubble water on the nanobubble mean diameter of N_2, O_2, $Ar + 8\%H_2$, air and CO_2 gas in deionized water (A), 1 mM NaCl aqueous solution (B), 1 mM $CaCl_2$ aqueous solution (C) and 1 mM $AlCl_3$ aqueous solution (D) are shown in Figure 3. The zeta potential of the nanobubbles and the pH and Eh as a function of time are shown in Figures 4 and 5, respectively. In this section, we introduce and discuss the results in the absence of salt followed by the results in the presence of salt. The Eh value change can be correlated to the H_2 nanobubble concentration change, O_2 nanobubble existence and CO_2 nanobubble dissolution in water.

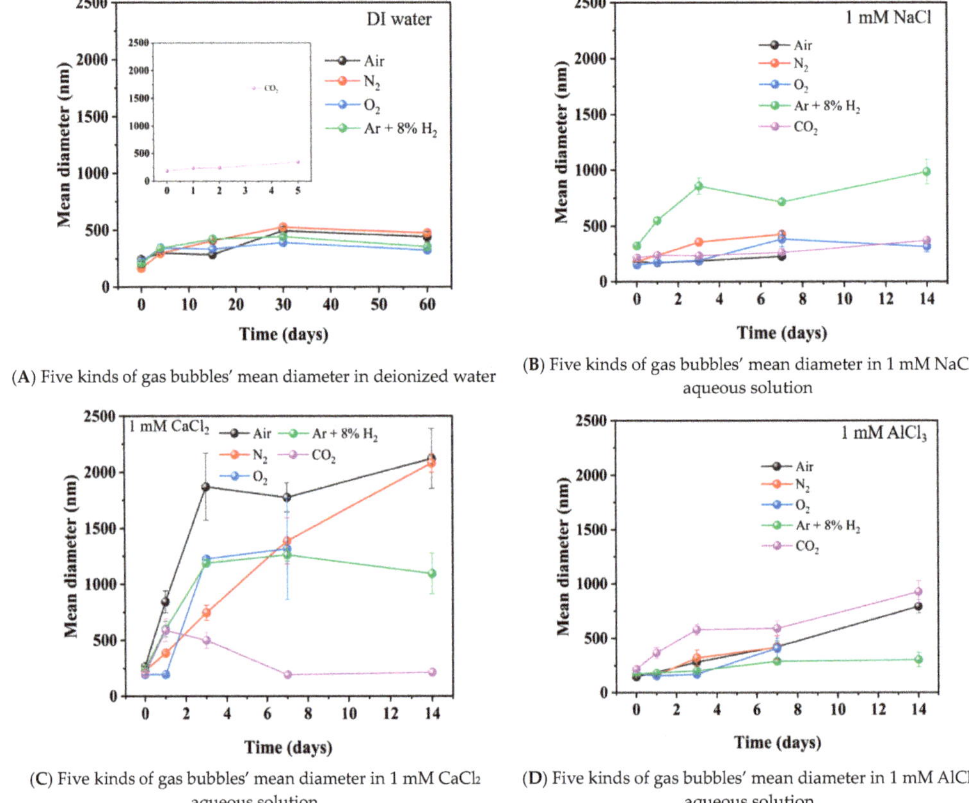

(**A**) Five kinds of gas bubbles' mean diameter in deionized water

(**B**) Five kinds of gas bubbles' mean diameter in 1 mM NaCl aqueous solution

(**C**) Five kinds of gas bubbles' mean diameter in 1 mM CaCl₂ aqueous solution

(**D**) Five kinds of gas bubbles' mean diameter in 1 mM AlCl₃ aqueous solution

Figure 3. Effect of time on nanobubble mean diameter of different kind of gases (N_2, O_2, Ar + 8%H_2 air and CO_2) in (**A**) deionized water, (**B**) 1 mM NaCl aqueous solution, (**C**) 1 mM CaCl₂ aqueous solution and (**D**) 1 mM AlCl₃ aqueous solution.

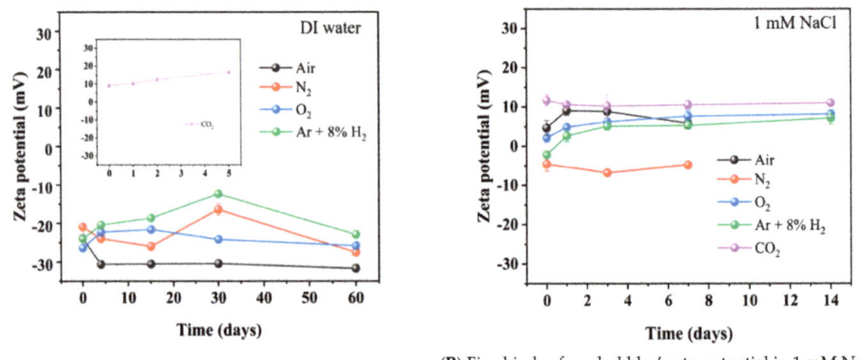

(**A**) Five kinds of gas bubbles' zeta potential in deionized water

(**B**) Five kinds of gas bubbles' zeta potential in 1 mM NaCl aqueous solution

Figure 4. *Cont.*

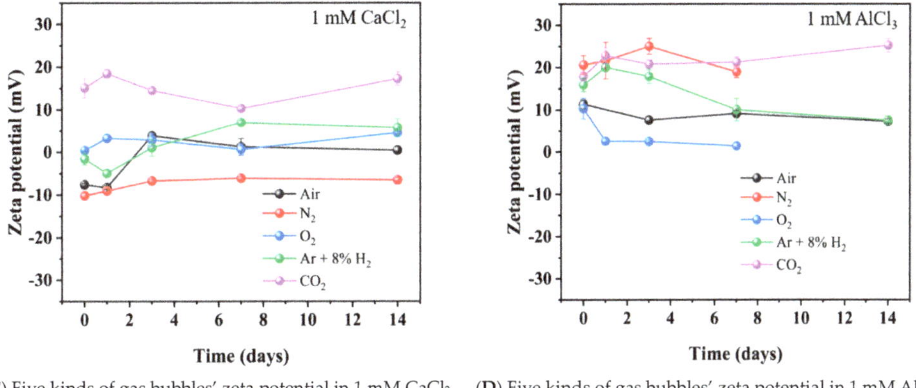

(**C**) Five kinds of gas bubbles' zeta potential in 1 mM CaCl₂ aqueous solution

(**D**) Five kinds of gas bubbles' zeta potential in 1 mM AlCl₃ aqueous solution

Figure 4. Effect of time on nanobubble zeta potential of different kind of gases (N_2, O_2, Ar + 8%H_2, air and CO_2) in (**A**) deionized water, (**B**) 1 mM NaCl aqueous solution, (**C**) 1 mM CaCl₂ aqueous solution and (**D**) 1 mM AlCl₃ aqueous solution.

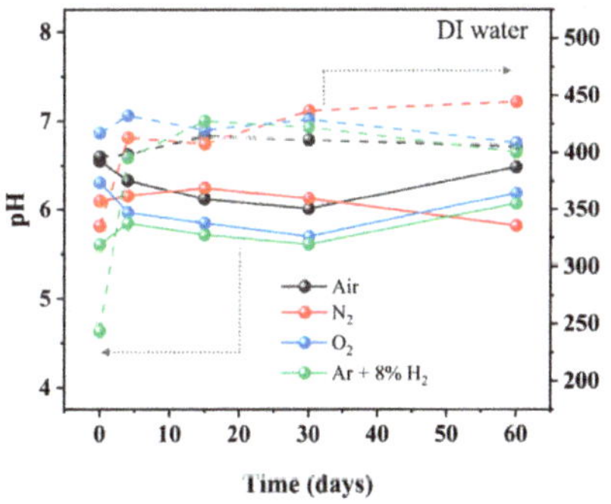

(**A**) Four kinds of gas bubbles' pH and Eh in deionized water

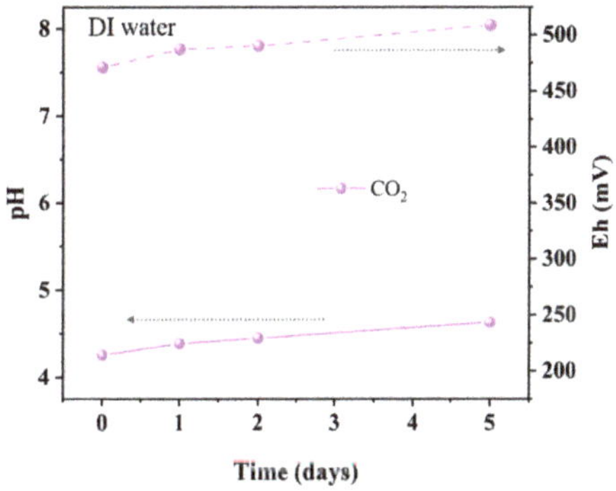

(**A′**) CO₂ gas bubbles' pH and Eh in deionized water

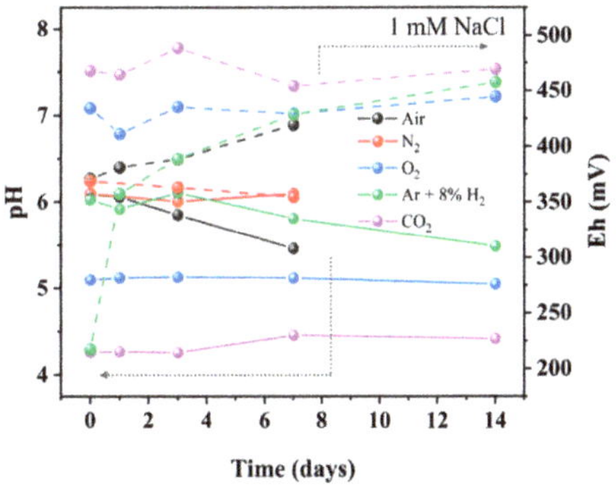

(**B**) Five kinds of gas bubbles' pH and Eh in 1 mM NaCl aqueous solution

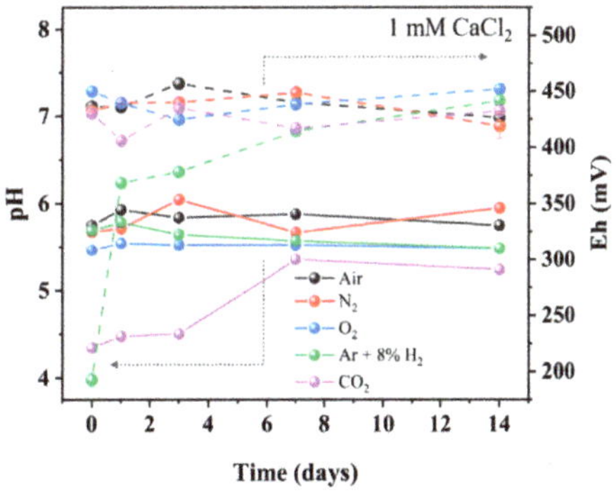

(**C**) Five kinds of gas bubbles' pH and Eh in 1 mM CaCl₂ aqueous solution

Figure 5. *Cont.*

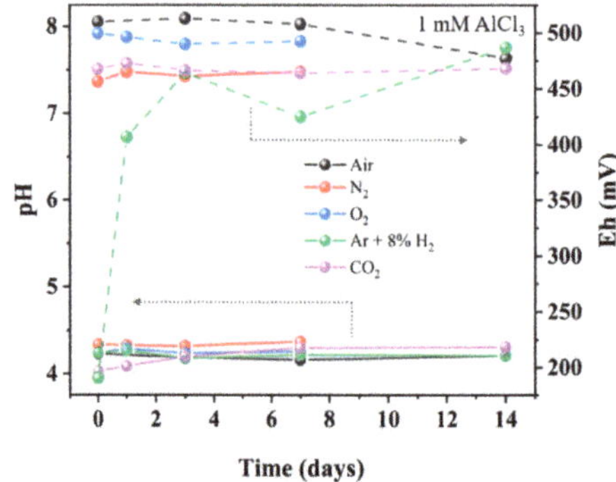

(**D**) Five kinds of gas bubbles' pH and Eh in 1 mM AlCl₃
aqueous solution

Figure 5. Effect of time on pH (solid line) and Eh (dashed line) of aqueous solution containing nanobubble of different kind of gases (N_2, O_2, Ar + 8%H_2, air and CO_2) in (**A,A′**) deionized water, (**B**) 1 mM NaCl aqueous solution, (**C**) 1 mM $CaCl_2$ aqueous solution and (**D**) 1 mM $AlCl_3$ aqueous solution.

3.2.1. Nanobubble Characteristics Change as a Function of Time in Deionized Water

In the absence of salt, the initial mean diameters of the N_2, O_2, Ar + 8%H_2 and air nanobubbles were around 200 nm and gradually increased to 400–530 nm in 30 days. After 30 days, the mean diameter slightly decreased and was stable at 330–480 nm in 60 days (Figure 3A). The zeta potentials of those gases were between −12 and −32 mV for 60 days (Figure 4A). Variation trends of nanobubbles are similar to previous reports. Ulatowski et al. (2019) reported the O_2 nanobubbles size produced by porous tube type was 200 nm on the initial day and unchanged after 10 days, while the N_2 nanobubble mean size was 300 nm at −11 mV of zeta potential on the initial day and 400 nm at about −12 mV after 35 days [3]. Meegoda et al. (2019) reported the O_2 nanobubble mean size produced by the cavitation method was 180 nm at −20 mV of zeta potential on the initial day and the O_2 bubble mean size increased to 285 nm at −16 mV after 7 days [9]. Michailidi et al. (2020) reported that with the nanobubbles prepared by the high shear cavitation method, initially the O_2 nanobubble mean size was 350 nm and the air nanobubble mean size was 430 nm, while the O_2 and air nanobubble sizes increased to 560 and 500 nm, respectively, after two weeks; both bubbles existed at 640 nm size after 2 and 3 months [4]. Although the nanobubble size depends on the preparation methods, the N_2 and O_2 nanobubbles' size increase with time was similar to our results that the prepared nanobubble sizes of various gases were about 200 nm initially and the nanobubble size increased to 300 to 500 nm with 2 to 4 weeks. Nirmalkar et al. (2018) reported that the number density of air nanobubbles prepared by the acoustic cavitation method gradually decreased; however, about 90 nm of nanobubbles existed for almost one year [6].

On the other hand, the CO_2 nanobubble size gradually increased with time and the size changed from 180 nm initially to 350 nm in 5 days (Figure 3A). After 5 days, the CO_2 nanobubbles were not stable and could not be detected by DLS because the CO_2 dissolved in the water and the electric double layer was compressed; therefore, the bubbles were coagulated due to the Van der Waals interaction (and the hydrophobic interaction) as discussed in the following Section 3.3.1. Ushikubo et al. (2010) reported that with the nanobubbles prepared by the pressurized type, the zeta potential of CO_2 nanobubbles was negative at pH 4, initially; however, the bubbles disappeared at pH 6.1 due to the dissociation of CO_2 [39]. In our experiment, the zeta potential of CO_2 nanobubbles was +9 mV at pH 4.2 on the first day and it gradually increased with time to +17 mV at pH 4.6 after 5 days (Figure 4A). As shown in Figure 5A, the Eh of Ar + 8%H_2 nanobubble suspension increased from 250 to 400 mV in one day and also the Eh of N_2 nanobubble suspension increased from 330 to 420 mV in one day. The Eh of O_2 and air nanobubble suspensions were between 400 and 450 mV initially, then to 60 days. The Eh of the CO_2 bubble suspension increased as the pH increased with time (Figure 5A′).

3.2.2. Nanobubble Characteristics Change as a Function of Time in Salt Aqueous Solutions

In the presence of salt, the nanobubbles were prepared in 1 mM NaCl, $CaCl_2$ or $AlCl_3$ aqueous solution. In 1 mM NaCl aqueous solution, N_2 and air nanobubble sizes were initially 200 nm and gradually increased; they were not observed after 7 days. Meanwhile, O_2, Ar + 8%H_2, and CO_2 bubble sizes were initially 200 to 300 nm, gradually increased and they were not detected after 14 days (Figure 3B). The existence period of CO_2 bubbles became longer in the 1 mM NaCl aqueous solution than in the absence of salt (i.e., 5 days in deionized water) (Figure 3A). The zeta potentials of N_2, O_2, Ar + 8%H_2 and air nanobubbles were small, within ±5 mV at initial days to 14 days (Figure 4B), while the zeta potential of CO_2 nanobubbles was higher than +12 mV initially and was kept +10 mV for 14 days (Figure 4B). Ke et al. (2019) reported that with the nanobubbles produced by the pressurized type, the initial N_2 nanobubble average size was about 200 nm in the 0.1 mM NaCl aqueous solution [2] and the size was similar to our result in the 1 mM NaCl condition (Figure 3B). In 1 mM NaCl aqueous solution, Leroy et al. (2012) measured that the zeta potential of H_2 gas was −20 mV at pH 6 [40] and Yang et al. (2001) reported the initial zeta potential of H_2 nanobubbles was −30 mV [41]. In our measurement of Ar + 8%H_2 nanobubble, the zeta potential was much smaller at −5 mV at pH 6 in 1 mM NaCl (Figure 3B). Meegoda et al. (2019) reported that the O_2 nanobubble mean size prepared by the high shear cavitation type was 214 nm at −22 mV of zeta potential initially and the O_2 bubble mean size was almost the same at 219 nm at −14 mV in 7 days [9]. In our experiment, the O_2 bubble mean size was about 300 nm for 14 days (Figure 3B), similar to the above-mentioned previous result; however, the zeta potential was +6 mV (Figure 4B). The Eh value increased from 220 to 360 mV with the Ar + 8%H_2 nanobubble suspension in one day, which was the same as nanobubbles in the deionized water suspension. The Eh of the N_2 nanobubble suspension was 360 mV and the Eh of other nanobubble suspensions was a little high at about 450 mV in 7 days (Figure 5B). The pH of CO_2 was from 4.3 to 4.4 and the Eh, pH, zeta potential and mean size of the CO_2 suspension were almost constant in the 1 mM NaCl aqueous solution compared with its suspension in deionized water.

In 1 mM $CaCl_2$ salt aqueous solution, O_2 was not stable after 7 days, while N_2, Ar + 8%H_2 and air bubbles could be observed until 14 days, but no more were detected after 14 days (Figure 3C). The CO_2 nanobubbles existed until about 20 days in 1 mM $CaCl_2$. The absolute value of the zeta potential of N_2, O_2, Ar + 8%H_2 and air nanobubbles was low, at less than 10 mV, while the zeta potential of CO_2 was higher than +10 mV (Figure 4C). Cho et al. (2005) reported that the air nanobubble size prepared by the sonicating method was 850 nm at −8 mV zeta potential [42], and their reported results were similar to our results after one day (Figures 3C and 4C). The size of the air, N_2 and O_2 nanobubbles in 1 mM $CaCl_2$ salt aqueous solution increased extremely with time. Yang et al. (2001) reported that the zeta potential of the H_2 nanobubble was −5.5 mV [8], and their result was similar to our result of Ar + 8%H_2 nanobubble after one day (Figure 4C). For the Eh value (Figure 5C), the Eh of the 8%H_2+Ar nanobubble suspension increased from 190 to 370 mV in one day, same as the nanobubbles in the other water suspensions. The Eh of N_2, O_2, air and CO_2 nanobubble suspensions were between 400 and 450 mV and the pH of N_2, O_2, Ar + 8%H_2 and air nanobubble suspensions were between 5.5 and 6. The pH of CO_2 solution increased from 4.4, initially, to 5.3 after 7 days (Figure 5C).

In 1 mM $AlCl_3$ salt aqueous solution, N_2 and O_2 were not stable after 7 days, while air, CO_2 and Ar + 8%H_2 nanobubbles could be observed until 14 days; however, they were not detected after 14 days. The size of the Ar + 8%H_2 nanobubble was more stable, at between 200 and 300 nm for 14 days, compared with other bubbles (Figure 3D). The zeta potentials of all prepared bubbles were positive (Figure 4D) possibly due to the presence of high valence ions (Al^{3+}) adsorbing on the bubble surfaces. Yang et al. (2001) reported that the zeta potential of H_2 nanobubble was +12 mV [41], and their result was similar to our result of +15 mV of Ar + 8%H_2 nanobubbles at the initial day (Figure 4D). Han et al. (2006) reported that the zeta potential of O_2 nanobubbles prepared by electrolysis was +20 mV at pH 6 in 10 mM NaCl + 1 mM $AlCl_3$ aqueous solution [43]. The trivalent ion Al^{3+} ion caused the positive charge in all our prepared nanobubbles (Figure 4D). For the Eh value (Figure 5D), the Eh of the Ar + 8%H_2 nanobubble suspension increased from 190 to 410 mV in one day, same as the gas nanobubbles in the other water suspensions. The Eh of N_2, O_2, air and CO_2 nanobubble suspensions were between 450 and 500 mV, and were higher than the Eh of the Ar + 8%H_2 suspension. The pH of N_2, O_2, Ar + 8%H_2 and air were from 4.15 to 4.35, which were lower than the pH of other suspensions. The pH of CO_2 slightly increased from 4.0 to 4.3 in 5 days (Figure 5A′) and its increase was limited compared with other suspensions in the salt aqueous solution and deionized water (Figure 5B).

The stable days of N_2, O_2, 8% H_2+Ar, air and CO_2 nanobubbles produced in different solutions are listed in Table 2. The N_2, O_2, Ar + 8%H_2 and air nanobubbles in 1 mM of the three salts with different valences (+1, +2 and +3) decreased the existence period for 1 to 2 weeks compared with deionized water in the absence of salt (>60 days). On the other hand, CO_2 bubbles existed only 5 days in deionized water in the absence of salt; however, CO_2 bubbles existed 14 days in the 1 mM salt aqueous solution. The stability differences will be further discussed in the following section.

Table 2. Stable days of gas nanobubbles produced in different solutions. The minimum days are given.

	DI Water	1 mM NaCl	1 mM CaCl$_2$	1 mM AlCl$_3$
Air	More than 60	7	14	14
N_2	More than 60	7	14	7
O_2	More than 60	14	7	7
Ar + 8%H_2	More than 60	14	14	14
CO_2	5	14	14	14

3.3. Nanobubble Stabilization Estimation by Using the Extended DLVO Theory

The thickness of the electric double layer (Debye length = $1/\kappa$) of the nanobubble is calculated in the next formula:

$$\kappa = \sqrt{\frac{2nz^2e^2}{\varepsilon_r\varepsilon_0 kT}} \tag{9}$$

where κ is the Debye–Hückel parameter, n is the concentration of anion or cation in the solution and is equal to 1000 $N_A C$ (N_A is the Avogadro number and C is concentration mol/L), z is the valence of ion, e is the electron charge, ε_r is the relative dielectric constant and ε_0 is the permittivity of vacuum.

The calculated Debye lengths of the nanobubbles studied in our experiment are listed in Table 3 in order to discuss differences in the nanobubble stability and to have some idea of how to keep the nanobubble stable for a longer duration. The Debye length is 300 nm at pH 6 around N_2, O_2 and Ar + 8%H_2 nanobubbles in deionized water, and 140 nm at pH 4.2 and 70 nm at pH 4.7 around CO_2 nanobubbles. Since the characteristics of CO_2 nanobubbles are different from others (Figures 3–5), the Debye length ($1/\kappa$) around CO_2 bubbles and HCO_3^-, OH^-, CO_3^{2-} ion concentration in the presence of CO_2 nanobubbles in deionized water as a function of pH were also calculated and shown in Figure 6 in order to discuss their relationship. The concentration of various ions was calculated from using Equations (3)–(8).

Table 3. Debye length ($1/\kappa$, nm) of nanobubbles prepared in aqueous solutions at natural pH.

	DI Water	1 mM NaCl	1 mM CaCl$_2$	1 mM AlCl$_3$
N$_2$	300 (pH 6)	10	5	3
O$_2$	300 (pH 6)	10	5	3
Ar + 8%H$_2$	300 (pH 6)	10	5	3
CO$_2$	140 (pH 4.2) 70 (pH 4.7)	10	5	3

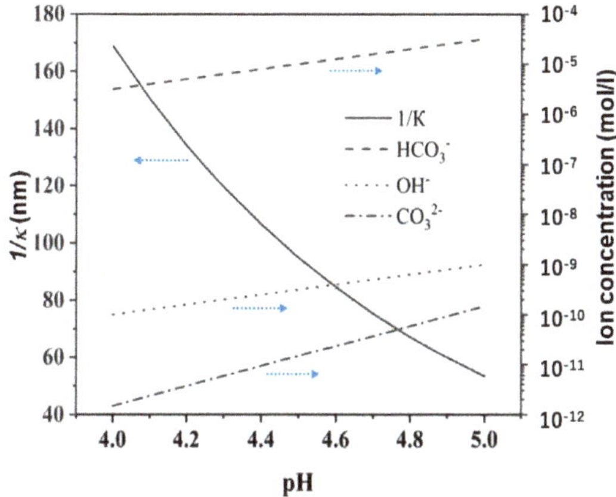

Figure 6. Debye length ($1/\kappa$) around CO$_2$ bubbles (left axis) and HCO$_3{}^-$, OH$^-$, CO$_3{}^{2-}$ ion concentration in the presence of CO$_2$ nanobubbles in deionized water (right axis) as a function of pH.

A thick electric double layer (300 nm, Table 3) is present around the N$_2$, O$_2$ and Ar + 8%H$_2$ nanobubbles in deionized water and can stabilize those nanobubbles with high enough zeta potential (i.e., -16 to -32 mV, Figure 4A). On the other hand, the Debye length of CO$_2$ nanobubbles (70 nm at pH 4.7 in DI water, Table 3) and other nanobubbles in salt aqueous solution (3–10 nm, Table 3) are thin and thus, the influence of the Van der Waals and hydrophobic attraction can be more dominant in coagulating and coalescing the nanobubbles. In order to quantify the influence of the electrical double layer potential, the Van der Waals potential and the hydrophobic interaction potential, their potential energies were calculated by using the extended DLVO theory as described in the following sections. Tchaliovska et al. investigated the thickness of thin flat foam films formed in aqueous dodecyl ammonium chloride solution [44]. Angarska et al. showed that the value of the critical thickness of the foam film was sensitive to the magnitude of the attractive surface forces acting on the film using sodium dodecyl sulfate and the total disjoining pressure operative on the films, which was expressed as a sum of the Van der Waals and hydrophobic contributions [45].

Figure 7 shows the two nanobubble positions and geometries considered in our potential calculation. The total potential energy V_T between two nanobubbles can be given by the potential energy due to the Van der Waals interaction (V_A), hydrophobic interaction (V_h) and the electrostatic interaction (V_R) as follows [46–50];

$$V_T = V_A + V_h + V_R, \quad \frac{V_T}{kT} = \frac{V_A + V_h + V_R}{kT} \tag{10}$$

$V_A + V_h$ is shown in the follow equation:

$$V_A + V_h = -\frac{A+K}{6}\left[\frac{2a_1 a_2}{R^2 - (a_1 + a_2)^2} + \frac{2a_1 a_2}{R^2 - (a_1 - a_2)^2} + \ln\left(\frac{R^2 - (a_1 + a_2)^2}{R^2 - (a_1 - a_2)^2}\right)\right] \tag{11}$$

where the Hamaker constant A for air in water whose value of air–water–air of 3.7×10^{-20} J [46] was used to investigate our system in nanobubble–water–nanobubble. The K is a hydrophobic constant for air in water. Yoon and Aksoy calculated the K using the extended DLVO theory [49] while Wang and Yoon measured the K as a function of the SDS concentration at different kinds of NaCl concentrations and showed that K was estimated at 10^{-17} J in the absence of salt and 10^{-19} J in the 1 mM NaCl aqueous solution [50]. In our calculation, 10^{-19} J of K was used in 1 mM $CaCl_2$ and $AlCl_3$ aqueous solutions and 10^{-18} J of K was used in deionized water containing CO_2 nanobubbles by considering the reference values [50].

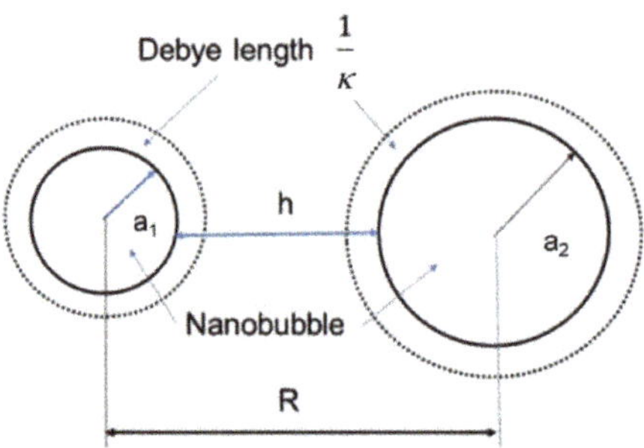

Figure 7. Position and geometry of two nanobubbles.

The radius of nanobubbles is a_1 and a_2 and their distance $R = a_1 + a_2 + h$ is defined as shown in Figure 7, and h is the surface-to-surface distance between two nanobubbles. The potential energy of V_R is expressed as follows,

$$V_R = -\frac{\pi \varepsilon_r \varepsilon_0 a_1 a_2 \left(\psi_1^2 + \psi_2^2\right)}{a_1 + a_2} \left[\frac{2\psi_1 \psi_2}{\psi_1^2 + \psi_2^2} ln \frac{1 + \exp(-\kappa h)}{1 - \exp(-\kappa h)} + \ln\{1 - \exp(-2\kappa h)\}\right] \quad (12)$$

where ψ_1 *and* ψ_2 are surface potentials of the nanobubbles of radii a_1 and a_2, respectively.

3.3.1. Stability Calculation of Nanobubbles in Deionized Water

The mean diameters of the prepared nanobubbles of N_2, O_2, Ar + 8%H_2 and air in deionized water were between 170 and 230 nm, as shown in Figure 3. With time, these diameters gradually increased, and they were between 390 and 530 nm after 30 days. Then, they slightly decreased to between 330 and 480 nm and were stable after 60 days. As the mean bubble diameter was measured in our experiment, the increase in bubble size indicates that the small size nanobubbles coalesce with other bubbles, and those small bubbles disappear.

The above-mentioned phenomenon is well known as the Ostwald ripening effect [51], which explains the deposition of a smaller object on a larger object due to the latter being more energetically stable than the former. By using Lemlich's theory [52], it can be explained by using next equation [53]:

$$\frac{da}{dt} = K' \left(\frac{1}{p} - \frac{1}{a}\right) \quad (13)$$

where a is the bubble radius, t is time, K' is a constant and p is instantaneous bubble mean radius. This equation tells us that the bubbles with a radius larger than a grow, and those with a radius smaller than p shrink. By using the results shown in Figure 2 (zeta potential) and Figure 3 (bubble size), the bubble stability is further discussed by calculating the total potential energy between two nanobubbles.

Figure 8 shows the potential energies between two nanobubbles in deionized water. Figure 8A shows the potential energies between two same-size gas bubbles of 200 nm of

Ar + 8%H$_2$ on the initial day. The total potential energy V_T is 20 kT at 300 nm (the Debye length) and the maximum potential is 30 kT. The other bubbles of air, N$_2$ and O$_2$ bubble show almost same potential curves. It indicates that the two bubbles have enough of a potential barrier to disperse each other. Figure 8B shows the potential energies between 450 nm of two same-size bubbles of Ar + 8%H$_2$ after 30 days. The maximum total potential energy V_T is 15 kT, i.e., the threshold total potential energy to determine coagulation or dispersion [54], and it indicates that Ar + 8%H$_2$ nanobubbles would be stable, although the absolute value of zeta potential is smallest in Ar + 8%H$_2$ nanobubbles (-13 mV, Figure 4A) compared with the ones of other O$_2$, N$_2$ and air nanobubbles bubbles, shown in Figure 4.

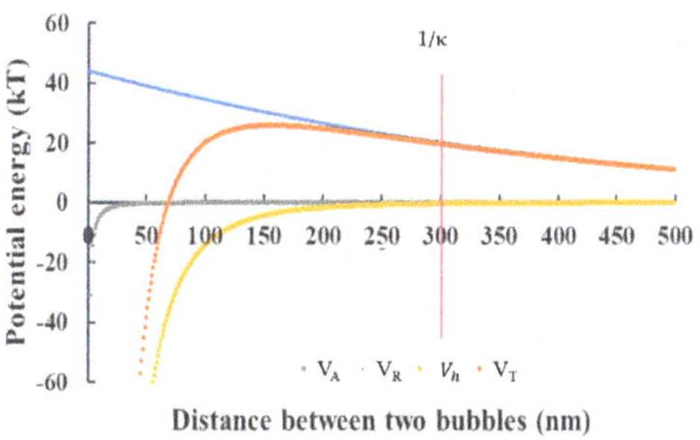

(**A**) Two 200 nm Ar + 8%H₂ gas bubbles on the prepared day

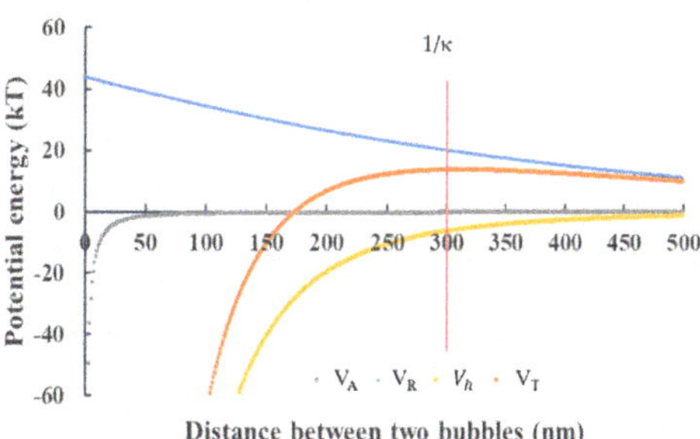

(**B**) Two 450 nm Ar + 8%H₂ gas bubbles after 30 days

Figure 8. Total potential energy V_T as a function of the distance between two nanobubbles at -24 mV zeta potential on the prepared day for Ar + 8%H$_2$ (**A**) and -12 mV after 30 days for Ar + 8%H$_2$ nanobubbles (**B**) in deionized water. (**A**) Two 200 Ar + 8%H$_2$ gas bubbles on the prepared day, (**B**) two 450 nm Ar + 8%H$_2$ gas bubbles after 30 days.

When the CO$_2$ nanobubbles were prepared, their mean size was 160 nm and zeta potential was low, i.e., +9 mV. Figure 9A shows the potential energies between two 160 nm bubbles. The total potential barrier was not appeared at 140 nm the Debye length. Figure 9B shows the potential energies between two 350 nm bubbles after 5 days. As the total potential barrier was also not appeared, these larger bubbles would also coagulate, coalesce and increase the size and, thus, CO$_2$ bubbles could not be observed after 5 days.

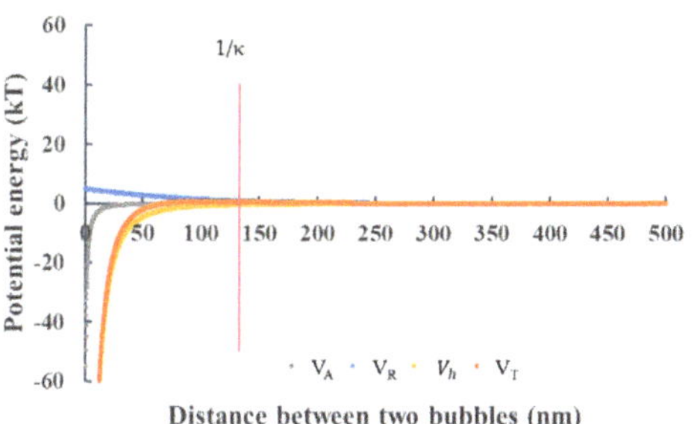

(**A**) Two 160 nm CO₂ gas bubbles on the prepared day

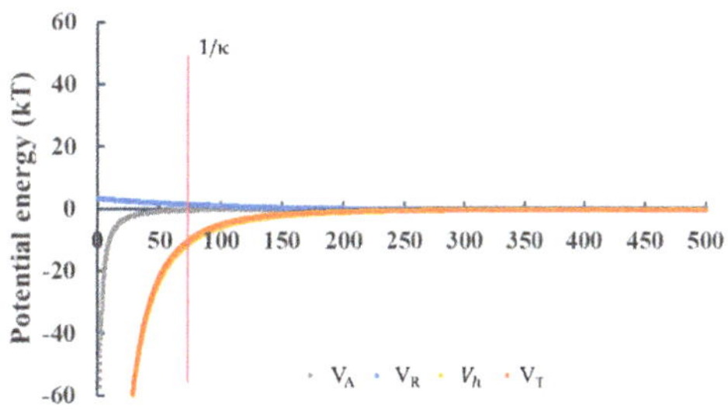

(**B**) Two 350 nm CO₂ gas bubbles after 5 days

Figure 9. Total potential energy V_T as a function of the distance between two nanobubbles at +9 mV zeta potential on the prepared day for CO$_2$ (**A**) and +5 mV after 5 days for CO$_2$ nanobubble (**B**) in deionized water. (**A**) Two 160 nm bubbles on the prepared day, (**B**) two 350 nm bubbles after 5 days.

3.3.2. Stability Calculation of Nanobubbles in Salt Aqueous Solutions

In the 1 mM NaCl aqueous solution, the absolute values of the zeta potential of O_2, N_2, Ar + 8%H_2 and air nanobubbles were less than 10 mV; the total potential energy barrier could not appear. Figure 10A shows the potential energies between two 160 nm N_2 nanobubbles. The total potential energy V_T is negative at any distance and does not show the potential barrier; therefore, the nanobubbles are not stable and can coagulate. Figure 10B shows the potential energies between two 200 nm CO_2 nanobubbles. Although the absolute zeta potential was slightly larger than the sN_2 nanobubbles at the Debye length 10 nm, there is no potential barrier indicating the attraction between the two bubbles. The CO_2 nanobubble size existed between 200 and 400 nm for 14 days, as shown in Figure 3B; however, after 14 days, the CO_2 bubble was disappeared.

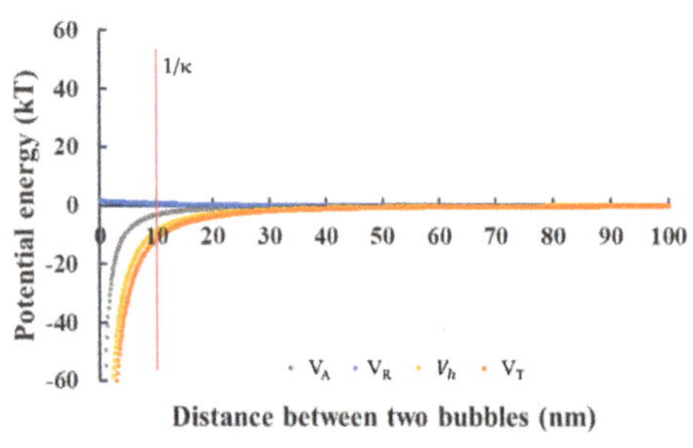

(**A**) Two 160 nm N_2 gas bubbles on the prepared day

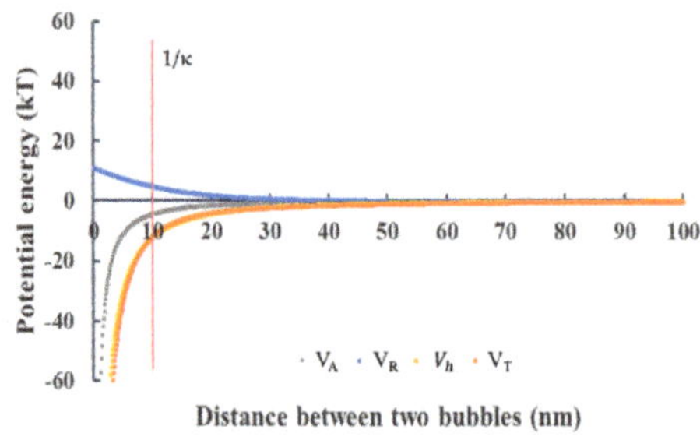

(**B**) Two 200 nm CO_2 gas bubbles on the prepared day

Figure 10. Total potential energy V_T as a function of the distance between two nanobubbles at −5 mV zeta potential on the prepared day for N_2 (**A**) and +12 mV on the prepared day for CO_2 nanobubble (**B**) in 1 mM NaCl aqueous solution. (**A**) Two 160 nm N_2 bubbles on the prepared day, (**B**) two 200 nm CO_2 bubbles on the prepared day.

In the 1 mM $CaCl_2$ aqueous solution, as the absolute values of the zeta potential of O_2, N_2, Ar + 8%H_2 and air nanobubbles were less than 10 mV (Figure 4C), like the one in 1 the mM NaCl aqueous solution (Figure 4B), the total potential energy barrier did not appear. Figure 11A shows the potential energies between 230 nm of two same-size N_2 nanobubbles on the prepared day. The total potential energy V_T was negative at any distance and did not show the potential barrier, and thus it indicates that the nanobubbles were not stable due to their coagulation. The zeta potential of CO_2 was more than +10 mV, as shown in Figure 4C; however, the potential barrier still did not appear because the Debye length around the CO_2 bubbles decreased in the presence of 1 mM $CaCl_2$ salt (i.e., 5 nm vs. 40 nm (pH 4.2) in deionized water, Table 2, while the CO_2 nanobubble mean size 1 mM $CaCl_2$ was stable at about 300 nm until 14 days (Figure 3C), same as the one in the 1 mM of NaCl aqueous solution (Figure 3B). Figure 11C shows the potential energies between two 750 nm N_2 nanobubbles after 3 days in a 1 mM $CaCl_2$ aqueous solution. With the smaller zeta potential −7 mV, the potential barrier is not appeared, and it corresponds to the further coagulation/coalescence of the bubbles (Figure 3C).

In the 1 mM $AlCl_3$ aqueous solution, Figure 11B shows the potential energy between two 170 nm Ar + 8%H_2 nanobubbles on the prepared day. Although the absolute value of the zeta potential was higher (+16 mV, Figure 4D) than the one of Ar + 8%H_2 nanobubbles prepared in a $CaCl_2$ aqueous solution (−2 mV, Figure 4C), there was no potential barrier due to the small electrostatic interaction potential (<15 kT) and the bubbles may be unstable. In Figure 11D, with the N_2 nanobubble size of 300 nm and the higher zeta potential (+25 mV, Figure 4D), the potential barrier was not also appeared.

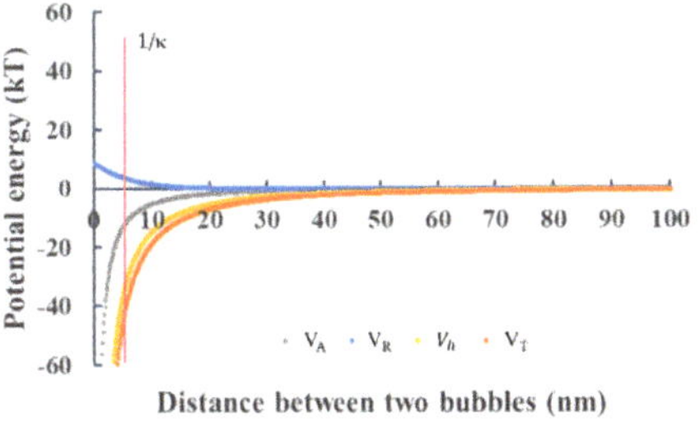

(**A**) Two 230 nm N_2 gas bubbles at −10 mV in 1 mM $CaCl_2$ aqueous solution on the prepared day

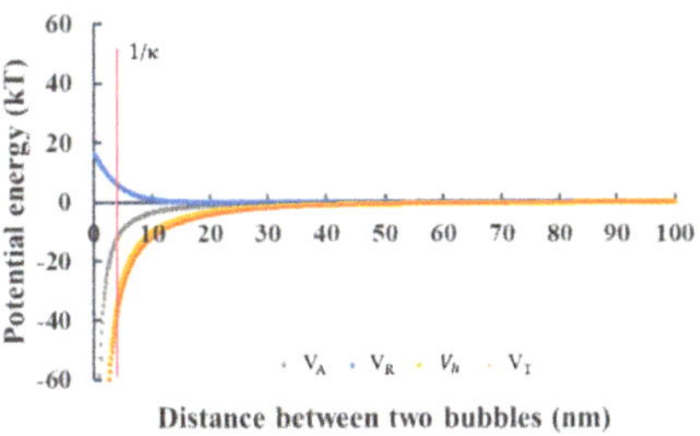

(**B**) Two 170 nm Ar + 8%H_2 gas bubbles at +16 mV in $AlCl_3$ aqueous solution on the prepared day

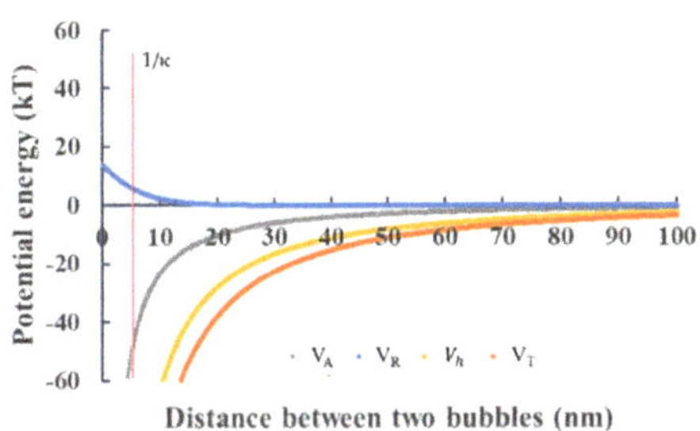

(**C**) Two 750 nm N_2 gas bubbles at -7mV in 1 mM $CaCl_2$ aqueous solution at 3rd day

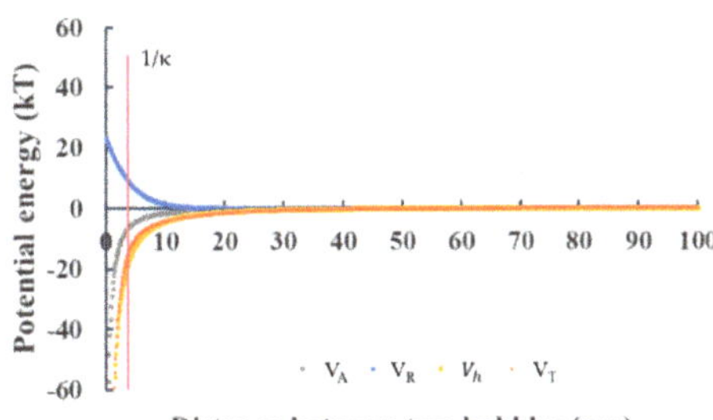

(**D**) Two 300 nm N_2 gas bubbles at +25 mV in 1 mM $AlCl_3$ aqueous solution after 3 days

Figure 11. Total potential energy V_T depending on the distance between two nanobubbles in the salt aqueous solution. (**A**) Two 230. nm N_2 nanobubbles at −10mV in 1 mM $CaCl_2$ aqueous solution on the prepared day, (**B**) Two 170 nm Ar + 8%H_2 nanobubbles at +16 mV in $AlCl_3$ aqueous solution on the prepared day, (**C**) Two 750 nm N_2 gas bubbles at −7mV in 1 mM $CaCl_2$ aqueous solution at 3rd day, (**D**) Two 300 nm N_2 nanobubbles at +25 mV in 1 mM $AlCl_3$ aqueous solution after 3 days.

In the case of N_2, O_2, Ar + 8%H_2 and air in 1 mM salt concentration, the presence of Ca^{2+} ion from $CaCl_2$ salt decreased the absolute value of the zeta potential by its adsorption on the bubble surface and a stronger coagulation could happen by thinning the electric double layer at the natural pH (without pH adjustment). Here, the overall bubble stability phenomena are summarized. Except for the 1 mM Ca^{2+} ion aqueous solution, the stability phenomena nearly followed the Shultz–Hardy rule. Comparing bubble size and existence time, the stability order of the N_2, O_2, Ar + 8%H_2 and air nanobubbles identified from our study was: no salt addition > 1 mM NaCl > 1 mM $AlCl_3$ > 1 mM $CaCl_2$ in deionized water. On the other hand, the stability order of CO_2 nanobubbles was: 1 mM NaCl > 1 mM $CaCl_2$ > 1 mM $AlCl_3$ > no salt addition, due to the effect of CO_2 dissolution in deionized water, as discussed in Section 3.3.

3.3.3. Nanobubble Movement in Salt Aqueous Solution

The nanobubbles move by Brownian motion. The Brownian diffusion can be described in the following Equations (14) and (15) where the bubble movement distance is $\sqrt{\overline{\Delta x^2}}$ and Equation (1) is incorporated to define D_T.

$$\overline{\Delta x^2} = 2D_T t = 2tkT / 3\pi \eta d_h \tag{14}$$

$$\sqrt{\overline{\Delta x^2}} = \sqrt{\frac{2tkT}{3\pi \eta d_h}} \tag{15}$$

If the bubble size d_h is small, the movement distance becomes longer since they have inverse correlation, as described in Equations (14) and (15).

On the other hand, a large, coalesced bubble floats, following the terminal velocity u defined by the Stokes equation of laminar flow:

$$u = \frac{d_h^2 \rho g}{18 \eta} \tag{16}$$

where ρ is the density of water and g is the gravitational acceleration.

Bubbles experiencing diffusion change their movement direction frequently in a short time. On the other hand, the bubble displacement due to the buoyancy force is always pointing upwards against the gravitational force. The terminal velocity depends on the bubble diameter, as shown in Figure 12, based on our calculation using Equation (16). A bubble smaller than 1 µm can experience very low terminal velocity ($<1 \times 10^{-6}$ m/s) which prevents bubbles from floating against the buoyancy force. Our results partially agree with the previous literature. Nirmalkar et al. (2018) reported that the number density of about a 100 nm mean size of nanobubbles gradually decreased over one year, and still existed at about 100 nm after one year [8].

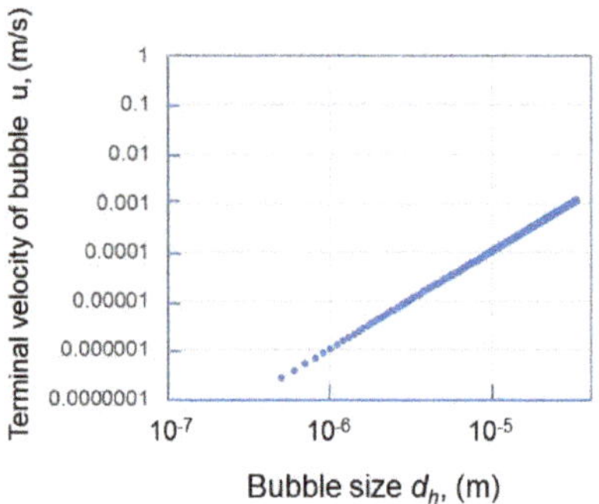

Figure 12. Relationship between the terminal velocity of a floating bubble and bubble size.

Figure 13 shows the schematic diagrams of nanobubble stability in deionized water (Figure 13A) and the nanobubble size change in a salt aqueous solution (Figure 13B). The nanobubbles in deionized water stably exist for two months by the Brownian motion. On the other hand, in the nanobubbles in a 1 mM salt aqueous solution, the nanobubbles coalesced with each other and increased the size, and the large-size bubbles gradually floated and disappeared after one or two weeks.

The surface charge on the front direction of a floating larger bubble can decrease, while the tail direction of the bubble retains more ions; thus, the electric double layer charge distribution can be distorted [55]. The bubbles in the direction of a floating large bubble can interact with that large bubble. If there is a small nanobubble in the front direction of a moving bubble, it can coagulate and coalesce with large bubbles and the coalesced large bubbles can be disappeared. This phenomenon occurred for the larger size of the bubbles in a salt aqueous solution with time, as observed in this study.

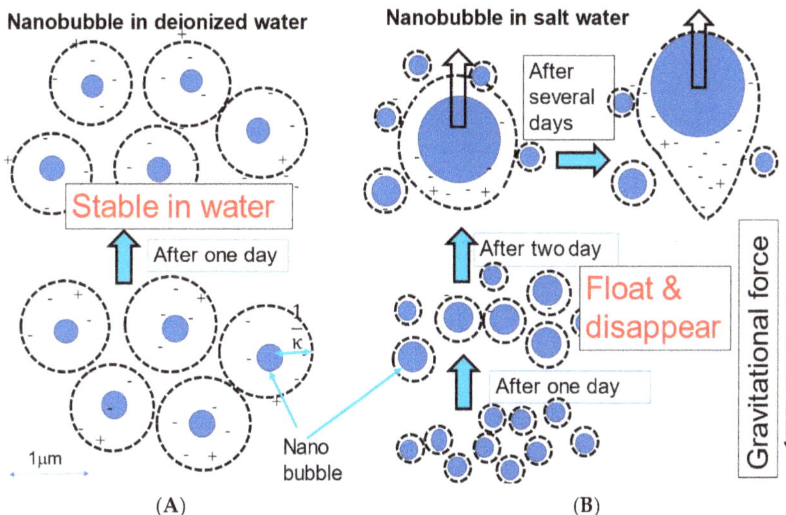

Figure 13. Schematic diagrams of nanobubble stability in deionized water (**A**) and nanobubble size change in salt aqueous solution (**B**) depending on time.

4. Conclusions

This research investigated five kinds of gas nanobubbles (N_2, O_2, Ar + 8%H_2, air and CO_2) for their long-term stability, to consider the application of the nanobubbles containing aqueous solutions. Each gas in a cylinder was injected into deionized water or a 1 mM of salt aqueous solution, and nanobubbles were prepared by the hydrodynamic cavitation method. The mean diameter, zeta potential of bubbles, pH and Eh of nanobubble suspensions were measured and these characteristics changed, as the function of time was also studied.

- The IEPs of different gas nanobubbles in deionized water varied, and the CO_2 nanobubbles showed the highest value of pH at 5.7.
- The N_2, O_2, Ar + 8%H_2 and air nanobubbles in deionized water showed the long-term stability for 60 days. During the 60 days, the bubble size gradually increased and decreased. Thus, this size change, explained by the Ostwald ripening effect, was also coupled with a bubble stability discussion using the total potential energy between two nanobubbles under different conditions calculated by the extended DLVO theory.
- The CO_2 nanobubbles produced in deionized water were not stable and disappeared after five days. The CO_2 nanobubbles in water dissolved the HCO_3^- ion, which could decrease the total potential energy between CO_2 bubbles, and thus the CO_2 nanobubbles became unstable.
- The N_2, O_2, Ar + 8%H_2 and air nanobubbles produced in the 1 mM salt aqueous solution were not stable. The potential barrier between the nanobubbles disappeared, and the bubble size gradually increased with their coalescence, followed by floating and disappearing after 14 days for O_2, Ar + 8%H_2 and air nanobubbles, and 7 days for N_2 nanobubbles. On the other hand, the CO_2 nanobubbles in the 1 mM of NaCl and $CaCl_2$ aqueous solution became more stable than the CO_2 nanobubbles in deionized water, and a 200 to 300 nm mean bubble size was kept for two weeks.
- A bubble smaller than 1 μm can experience very low terminal velocity ($<1 \times 10^{-6}$ m/s) which prevents bubbles from floating against the buoyancy force.

Author Contributions: Y.Z.: Conceptualization, data curation, writing; Z.H.: Data curation; C.H.: Funding acquisition, Investigation; K.W.: Formal analysis; Y.W. (Youbin Wang): Formal analysis; N.L. and Q.F.: Methodology; G.D. and A.O.: Writing and editing; Y.W. (Yuezhou Wei): Supervision, investigation; T.F.: Conceptualization, supervision, writing, reviewing and editing. All authors have read and agreed to the published version of the manuscript.

Funding: One part of this research was funded by the Natural Science Foundation of China, grant number NSFC 21976039.

Institutional Review Board Statement: Not applicable.

Informed Consent Statement: Not applicable.

Data Availability Statement: The authors confirm that the data supporting the findings of this study are available within the article.

Acknowledgments: The authors appreciate the reviewers.

Conflicts of Interest: The authors declare no competing interests.

Nomenclature

a	Radius of nanobubble
a_1	Radius of nanobubble 1
a_2	Radius of nanobubble 2
A	Hamaker constant for the gas in water
C	Concentration
d_h	Hydrodynamic diameter
D_T	Diffusion coefficient
e	Electron charge
g	Gravitational acceleration
h	surface-to-surface distance between two nanobubbles
k	Boltzmann constant
K_{a1}, K_{a2}	Equilibrium constant
K	hydrophobic force constant for the gas in water
K'	constant for Lemlich's theory
n	Concentration of anion or cation in solution
N_A	Avogadro number
ΔP	Laplace pressure difference
p	Instantaneous bubble mean radius
R	Distance between bubbles 1 and 2
t	Time
T	Absolute temperature
u	Terminal velocity
V_A	Van der Waals interaction potential energy
V_h	Hydrophobic force interaction potential energy
V_R	Electrostatic interaction potential energy
V_T	Total potential energy of interaction between two nanobubbles
x	Distance
z	Ionic valence
γ	Surface tension
ε	Electron charge
ε_0	Permittivity of vacuum
ε_r	Relative dielectric constant
Ψ_1, Ψ_2	Surface potential of the nanobubbles of radius a_1 and a_2
η	Viscosity
κ	Debye–Hückel parameter
ρ	Density of water
μ	Electrophoretic mobility
ζ	Zeta potential

References

1. Xiao, W.; Xu, G. Mass transfer of nanobubble aeration and its effect on biofilm growth: Microbial activity and structural properties. *Sci. Total Environ.* **2020**, *703*, 134976. [CrossRef] [PubMed]
2. Ke, S.; Xiao, W.; Quan, N.; Dong, Y.; Zhang, L.; Hu, J. Formation and Stability of Bulk Nanobubbles in Different Solutions. *Langmuir* **2019**, *35*, 5250–5256. [CrossRef]
3. Ulatowski, K.; Sobieszuk, P.; Mróz, A.; Ciach, T. Stability of nanobubbles generated in water using porous membrane system. *Chem. Eng. Process. Process Intensif.* **2019**, *136*, 62–71. [CrossRef]
4. Michailidi, E.D.; Bomis, G.; Varoutoglou, A.; Kyzas, G.Z.; Mitrikas, G.; Mitropoulos, A.C.; Efthimiadou, E.K.; Favvas, E.P. Bulk nanobubbles: Production and investigation of their formation/stability mechanism. *J. Colloid Interface Sci.* **2020**, *564*, 371–380. [CrossRef] [PubMed]
5. Kanematsu, W.; Tuziuti, T.; Yasui, K. The influence of storage conditions and container materials on the long term stability of bulk nanobubbles -Consideration from a perspective of interactions between bubbles and surroundings. *Chem. Eng. Sci.* **2020**, *219*, 115594. [CrossRef]
6. Nirmalkar, N.; Pacek, A.W.; Barigou, M. On the Existence and Stability of Bulk Nanobubbles. *Langmuir* **2018**, *34*, 10964–10973. [CrossRef]
7. Uchida, T.; Liu, S.; Enari, M.; Oshita, S.; Yamazaki, K.; Gohara, K. Effect of NaCl on the Lifetime of Micro- and Nanobubbles. *Nanomaterials* **2016**, *6*, 31. [CrossRef]
8. Nirmalkar, N.; Pacek, A.W.; Barigou, M. Interpreting the interfacial and colloidal stability of bulk nanobubbles. *Soft Matter* **2018**, *14*, 9643–9656. [CrossRef] [PubMed]
9. Meegoda, J.N.; Hewage, S.A.; Batagoda, J.H. Application of the diffused double layer theory to nanobubbles. *Langmuir* **2019**, *39*, 12100–12112. [CrossRef]
10. Hewage, S.A.; Kewalramani, J.; Meegoda, J.N. Stability of nanobubbles in different salt solutions. *Colloids Surf. A* **2021**, *609*, 125669. [CrossRef]
11. Fujita, T.; Kurokawa, H.; Han, Z.; Zhou, Y.; Matsui, H.; Ponou, J.; Dodbiba, G.; He, C.; Wei, Y. Free Radical Degradation in Aqueous Solution by BlowingHydrogen and Carbon Dioxyde Nanotubes. *Sci. Rep.* **2021**, *11*, 3068. [CrossRef] [PubMed]
12. Kurokawa, H.; Ito, H.; Taninaka, A.; Shigekawa, H.; Dodbiba, G.; Wei, Y.; Fujita, T. Antioxidant Effect of Hydrogen Nanobubble Contributes to Suppression of Tumor Cell Growth. *Biomed. J. Sci. Tech. Res.* **2019**, *19*, 14592–14594.
13. Temesgen, T.; Bui, T.T.; Han, M.; Kim, T.-I.; Park, H. Micro and nanobubble technologies as a new horizon for water-treatment techniques: A review. *Adv. Colloid Interface Sci.* **2017**, *246*, 40–51. [CrossRef]
14. Wu, Y.; Lyu, T.; Yue, B.; Tonoli, E.; Verderio, E.A.M.; Ma, Y.; Pan, G. Enhancement of Tomato Plant Growth and Productivity in Organic Farming by Agri-Nanotechnology Using Nanobubble Oxygation. *J. Agric. Food Chem.* **2019**, *67*, 10823–10831. [CrossRef]
15. Ghadimkhani, A.; Zhang, W.; Marhaba, T. Ceramic membrane defouling (cleaning) by air Nano Bubbles. *Chemosphere* **2016**, *146*, 379–384. [CrossRef] [PubMed]
16. Zhu, J.; An, H.; Alheshibri, M.; Liu, L.; Terpstra, P.M.; Liu, G.; Craig, V.S. Cleaning with Bulk Nanobubbles. *Langmuir* **2016**, *32*, 11203–11211. [CrossRef] [PubMed]
17. Dayarathne, H.N.P.; Jeong, S.; Jang, A. Chemical-free scale inhibition method for seawater reverse osmosis membrane process: Air micro-nano bubbles. *Desalination* **2019**, *461*, 1–9. [CrossRef]
18. Atkinson, A.J.; Apul, O.G.; Schneider, O.; Garcia-Segura, S.; Westerhoff, P. Nanobubble Technologies Offer Opportunities to Improve Water Treatment. *Acc. Chem. Res.* **2019**, *52*, 1196–1205. [CrossRef]
19. Azevedo, A.; Oliveira, H.A.; Rubio, J. Treatment and water reuse of lead-zinc sulphide ore mill wastewaters by high rate dissolved air flotation. *Miner. Eng.* **2018**, *127*, 114–121. [CrossRef]
20. Kyzas, G.Z.; Bomis, G.; Kosheleva, R.I.; Efthimiadou, E.K.; Favvas, E.P.; Kostoglou, M.; Mitropoulos, A.C. Nanobubbles effect on heavy metal ions adsorption by activated carbon. *Chem. Eng. J.* **2019**, *356*, 91–97. [CrossRef]
21. Lyu, T.; Wu, S.; Mortimer, R.J.G.; Pan, G. Nanobubble Technology in Environmental Engineering: Revolutionization Potential and Challenges. *Environ. Sci. Technol.* **2019**, *53*, 7175–7176. [CrossRef]
22. Agarwal, A.; Ng, W.J.; Liu, Y. Principle and applications of microbubble and nanobubble technology for water treatment. *Chemosphere* **2011**, *84*, 1175–1180. [CrossRef]
23. Hernandez, C.; Abenojar, E.C.; Hadley, J.; de Leon, A.C.; Coyne, R.; Perera, R.; Gopalakrishnan, R.; Basilion, J.P.; Kolios, M.C.; Exner, A.A. Sink or float? Characterization of shell-stabilized bulk nanobubbles using a resonant mass measurement technique. *Nanoscale* **2019**, *11*, 851–855. [CrossRef]
24. Azevedo, A.; Etchepare, R.; Calgaroto, S.; Rubio, J. Aqueous dispersions of nanobubbles: Generation, properties and features. *Miner. Eng.* **2016**, *94*, 29–37. [CrossRef]
25. Etchepare, R.; Azevedo, A.; Calgaroto, S.; Rubio, J. Removal of ferric hydroxide by flotation with micro and nanobubbles. *Sep. Purif. Technol.* **2017**, *184*, 347–353. [CrossRef]
26. Yang, X.; Nie, J.; Wang, D.; Zhao, Z.; Kobayashi, M.; Adachi, Y.; Shimizu, K.; Lei, Z.; Zhang, Z. Enhanced hydrolysis of waste activated sludge for methane production via anaerobic digestion under N2-nanobubble water addition. *Sci. Total Environ.* **2019**, *693*, 133524. [CrossRef]
27. Wang, X.; Lei, Z.; Shimizu, K.; Zhang, Z.; Lee, D.J. Improved methane production from corn straw using anaerobically digested sludge pre-augmented by nanobubble water. *Bioresour. Technol.* **2020**, *311*, 123479. [CrossRef] [PubMed]

28. Phan, K.K.T.; Truong, T.; Wang, Y.; Bhandari, B. Nanobubbles: Fundamental characteristics and applications in food processing. *Trends Food Sci. Technol.* **2020**, *95*, 118–130. [CrossRef]
29. Phan, K.K.T.; Truong, T.; Wang, Y.; Bhandari, B. Formation and Stability of Carbon Dioxide Nanobubbles for Potential Applications in Food Processing. *Food Eng. Rev.* **2020**. [CrossRef]
30. Kim, H.J.; Choi, H.; Choi, H.; Lee, B.; Lee, D.; Lee, D.E. Study on Physical Properties of Mortar for Section Restoration Using Calcium Nitrite and CO_2 Nano-Bubble Water. *Materials* **2020**, *13*, 3897. [CrossRef] [PubMed]
31. Kim, J.; Kitagaki, R.; Choi, H. Pore Filling Effect of Forced Carbonation Reactions Using Carbon Dioxide Nanobubbles. *Materials* **2020**, *13*, 4343. [CrossRef]
32. Oh, S.H.; Yoon, S.H.; Song, H.; Han, J.G.; Kim, J.-M. Effect of hydrogen nanobubble addition on combustion characteristics of gasoline engine. *Int. J. Hydrog. Energy* **2013**, *38*, 14849–14853. [CrossRef]
33. Wang, X.; Yuan, T.; Lei, Z.; Kobayashi, M.; Adachi, Y.; Shimizua, K.; Lee, D.J.; Zhang, Z. Supplementation of O-2-containing gas nanobubble water to enhance methane production from anaerobic digestion of cellulose. *Chem. Eng. J.* **2020**, *398*, 125652. [CrossRef]
34. Trefalt, G.; Szilagyi, I.; Téllez, G.; Borkovec, M. Colloidal Stability in Asymmetric Electrolytes: Modifications of the Schulze-Hardy Rule. *Langmuir* **2017**, *33*, 1695–1704. [CrossRef]
35. Reeks, M.W. *Stokes-Einstein Equation*; Begell House: Danbury, CT, USA, 2011.
36. Oshima, H. Developments in Theories of Electrokinetic Phenomena: From Smoluchowski to ELKIN. *Oleoscience* **2013**, *13*, 3–9.
37. Lide, D.R. *Handbook of Chemistry and Pysics*, 101st ed.; CRC Press: Boca Raton, FL, USA, 2020.
38. Welch, M.J.; Lipton, J.F.; Seck, J.A. Tracer studies with radioactive oxygen-15. Exchange between carbon dioxide and water. *J. Phys. Chem.* **1969**, *73*, 351–3356. [CrossRef]
39. Ushikubo, F.Y.; Enari, M.; Furukawa, T.; Nakagawa, R.; Makino, Y.; Kaeagoe, Y.; Oshita, S. Zeta-potential of Micro- and/or Nano-bubbles in Water Produced by some kinds of Gases. *IFAC Proc.* **2010**, *43*, 283–288. [CrossRef]
40. Leroy, P.; Dougnot, D.; Revil, A.; Lassin, A.; Azaroual, M. A double layer model of the gas bubble/water interface. *J. Colloid Interface Sci.* **2012**, *388*, 243–256. [CrossRef]
41. Yang, C.; Dabros, T.; Li, D.; Czarnecki, J.; Masliyah, J.H. Measurement of the zeta potential of gas bubble in aqueous solutions by micro electrophoresis method. *J. Colloid Interface Sci.* **2001**, *243*, 128–135. [CrossRef]
42. Cho, S.-H.; Kim, J.-Y.; Chun, J.-H.; Kim, J.-D. Ultrasonic formation of nanobubbles and their zeta-potentials in aqueous electrolyte and surfactant solutions, Colloids and Surfaces A: Physicochem. *Eng. Asp.* **2005**, *269*, 28–34. [CrossRef]
43. Han, M.Y.; Kim, M.K.; Shin, M.S. Generation of a positively charged bubble and its possiblemechanism of formation. *J. Water Supply Res. Technol. AQUA* **2006**, *55*, 471–478. [CrossRef]
44. Tchaliovska, S.; Manev, E.; Radoev, B.; Eriksson, J.C.; Claesson, P.M. Interactions in equiribrium free films of aqueous dodecyl ammonium chloride solutions. *J. Colloid Interface Sci.* **1994**, *168*, 190–1944. [CrossRef]
45. Angarska, J.K.; Dimitrova, B.S.; Danov, K.D.; Kralchevsky, P.A.; Ananthapadmanabhan, K.P.; Lips, A. Detection of the hydrophobic surface force in foam films by measurements of the critical thickness of the film rupture. *Langmuir* **2004**, *20*, 1799–1806. [CrossRef]
46. Israelachivili, J.N. *Intermolecular and Surface Forces with Applications to Colloidal and Biological Systems*; Academic Press Limited: London, UK, 1985.
47. Usui, S.; Sasaki, H. Dispersion and Coagulation of Fine Particles—Fundamentals and applications. *J. Shigen Sozai* **1991**, *107*, 585–591. [CrossRef]
48. Fujita, T.; Ito, R.; Tokoro, C.; Sadaki, J.; Dodbiba, G.; Tsukamoto, R.; Okuda, H.; Yamane, H. Classification of submicron Ni particles by heterocoagulation. *Powder Technol.* **2007**, *173*, 19–28. [CrossRef]
49. Yoon, R.H.; Aksoy, B.S. Hydrophobic forces in thin water films stabilized by dodecyl ammonium chloride. *J. Colloid Interface Sci.* **1999**, *211*, 1–10. [CrossRef] [PubMed]
50. Wang, L.; Yoon, R.H. hydrophobic forces n the foam films stabilized by sodium dodecyl sulfate: Effect of electrolyte. *Langmuir* **2004**, *20*, 11457–11464. [CrossRef]
51. Zheng, X.; Jang, J. Hydraulic properties of porous media saturated with nanoparticle-stabilized air-water foam. *Sustainability* **2016**, *8*, 1317. [CrossRef]
52. Lemilich, R. Prediction of changes in bubble size distribution due to interbubble gas distribution due to interbubble gas diffusion in form. *Ind. Eng. Chem. Fundam.* **1978**, *17*, 89–93. [CrossRef]
53. Huang, Z.; Su, M.; Yang, Q.; Li, Z.; Chen, S.; Li, Y.; Zhou, X.; Li, F.; Song, Y. A general patterning approach by manipulating the evolution of two-dimensional liquid foams. *Nat. Commun.* **2017**, *8*, 14110. [CrossRef]
54. Otsuki, A.; Bryant, G. Characterization of the interactions within fine particle mixtures in highly concentrated suspensions for advanced particle processing. *Adv. Colloid Interface Sci.* **2015**, *226*, 37–43. [CrossRef] [PubMed]
55. Derjaguin, B.W.; Dukhin, S.S. Theory of flotation of small and medium size particles. *Trans. Inst. Min. Metall.* **1960**, *70*, 221–246. [CrossRef]

Article

The Influence of Air Nanobubbles on Controlling the Synthesis of Calcium Carbonate Crystals

Yongxiang Wu [1], Minyi Huang [1], Chunlin He [1], Kaituo Wang [1], Nguyen Thi Hong Nhung [1], Siming Lu [2], Gjergj Dodbiba [3], Akira Otsuki [4,5] and Toyohisa Fujita [1,*]

[1] School of Resources, Environment and Materials, Guangxi University, Nanning 530004, China
[2] Zhengzhou Non-Ferrous Metals Research Institute Ltd. of CHALCO, Zhengzhou 450041, China
[3] Graduate School of Engineering, The University of Tokyo, Tokyo 113-8656, Japan
[4] Facultad de Ingeniería y Ciencias, Universidad Adolfo Ibáñez, Diagonal Las Torres 2640, Peñalolén, Santiago 7941169, Chile
[5] Waste Science & Technology, Luleå University of Technology, SE 971 87 Luleå, Sweden
*   Correspondence: fujitatoyohisa@gxu.edu.cn

Abstract: Numerous approaches have been developed to control the crystalline and morphology of calcium carbonate. In this paper, nanobubbles were studied as a novel aid for the structure transition from vaterite to calcite. The vaterite particles turned into calcite (100%) in deionized water containing nanobubbles generated by high-speed shearing after 4 h, in comparison to a mixture of vaterite (33.6%) and calcite (66.3%) by the reaction in the deionized water in the absence of nanobubbles. The nanobubbles can coagulate with calcite based on the potential energy calculated and confirmed by the extended DLVO (Derjaguin–Landau–Verwey–Overbeek) theory. According to the nanobubble bridging capillary force, nanobubbles were identified as the binder in strengthening the coagulation between calcite and vaterite and accelerated the transformation from vaterite to calcite.

Keywords: nanobubbles; calcium carbonate; transformation; crystal-control; extended DLVO theory; capillary force

Citation: Wu, Y.; Huang, M.; He, C.; Wang, K.; Nhung, N.T.H.; Lu, S.; Dodbiba, G.; Otsuki, A.; Fujita, T. The Influence of Air Nanobubbles on Controlling the Synthesis of Calcium Carbonate Crystals. *Materials* 2022, *15*, 7437. https://doi.org/10.3390/ma15217437

Academic Editor: Carlos Javier Duran-Valle

Received: 2 October 2022
Accepted: 20 October 2022
Published: 23 October 2022

Publisher's Note: MDPI stays neutral with regard to jurisdictional claims in published maps and institutional affiliations.

Copyright: © 2022 by the authors. Licensee MDPI, Basel, Switzerland. This article is an open access article distributed under the terms and conditions of the Creative Commons Attribution (CC BY) license (https://creativecommons.org/licenses/by/4.0/).

1. Introduction

Nanobubbles are described as a gaseous domain in the liquid phase with a diameter of less than 1 μm [1] with unusual longevity lasting for several weeks or even months, which contradicts the classical Epstein–Plesset theory prediction that nanobubbles cannot exist and should disappear in a few milliseconds or microseconds [2,3]. Until now, significant progress has been made on the stability of nanobubbles [2–4], which has attracted the attention of experimentalists and theorists. There are more and more studies involving possible explanations for the stability of nanobubbles that have been proposed [5]. Since it was discovered that the surface of nanobubbles had a high magnitude of zeta potential, the charge stabilization model has aroused the attention of researchers [6–8]. In recent years, the mechanism of charge stabilization appears to be the most reasonable to rationalize nanobubble stabilization. An electrical double layer is formed on the surface of the nanobubble in water containing free moving ions, which serves as the main energy supply to withstand the coalescence of multiple bubbles and generates a repulsive pressure to balance the Laplace pressure [2].

The majority of nanobubble research has recently been rapidly increasing in many kinds of scientific disciplines. The unique physicochemical properties of the nanobubbles, such as long-term stability [2,3,6], high zeta potential [3,7], generation, and degradation of free radicals when collapsing [9], has aroused interest in many areas. Therefore, numerous reports have been published on the application of nanobubbles, including medicine [10,11], environment [9,12], flotation [13,14], and materials [15]. In medicine, nanobubbles have been used to protect proteins from surface-mediated denaturation [16] as well as for ultrasound imaging and intracellular drug delivery [17]. H_2 nanobubbles in water were discovered to inhibit tumor cell growth [18]. The application of nanobubbles in environmental studies is well-known, such as the removal of heavy metal ions [19], wastewater treatment [20], removal of pollutants from water [21]. As mentioned above, nanobubbles are thought to have a wide range of applications. However, the application of nanobubbles in material science and engineering is uncommon, necessitating further exploration of their potential applications.

To some extent, nanobubbles can change the state and properties of water, as mentioned in many papers. Hydrogen nanobubbles in aqueous solution have a very low Eh (oxidation and reduction potential), which can be used to prevent oxidation [8]. Because a large number of hydroxyl radicals are generated when oxygen nanobubbles in aqueous solution break, they have a high oxidation state [8]. Due to their unique physicochemical properties, in this research, nanobubbles in water were used as a reaction solvent instead of pure water during calcium carbonate ($CaCO_3$) synthesis by the double decomposition method to study the influence of nanobubbles on the growth of crystalline $CaCO_3$.

Calcium carbonate ($CaCO_3$), one of nature's most common and widely dispersed materials, is an important component of the global carbon cycle [22]. Calcium carbonate has three crystalline phases vaterite, aragonite, and calcite [23]. In these crystals, vaterite is the least stable phase in the water, and can spontaneously change into calcite [24]. Calcite is the most common form because it is thermodynamically stable [23]. The formation of vaterite to calcite can be influenced by reaction parameters, such as supersaturation, temperature, reaction time, and additive use [25].

Furthermore, calcium carbonate has a wide range of applications that depend on its shape and morphology. Therefore, it is important to control the crystal shape or morphology of $CaCO_3$ [26]. Aragonite whiskers with high aspect ratios, for example, have been in great demand for the reinforcement of polymer materials [26]. Flexible and deformable calcium carbonate is also extensively used as a filler and pigment in papermaking [27]. Surface-functionalized calcium carbonate particles can also be adapted to create novel catalytic materials [28]. Non-spherical vaterite particles are appealing as solid supports for targeted and extended drug delivery [29]. In wastewater treatment, vaterite has been explored as it is excellent at removing several heavy metal ions [30–32]. Porous vaterite and cubic calcite aggregated calcium carbonate are used for Cu^{2+} heavy metal removal [32]. Vaterite particles for strontium removal were also investigated and reported [31]. However, Sasamoto reported that porous calcite has potential as an adsorbent with a fast reaction rate and large adsorption amount in the removal of heavy metal ions [33].

There are several preparation methods for $CaCO_3$, including the carbonation method [34], double decomposition method [35], and the thermal decomposition of calcium bicarbonate [36]. Among them, the double decomposition method is a common way to produce $CaCO_3$. In this method, calcium chloride ($CaCl_2$) is exploited as a calcium source, and sodium carbonate (Na_2CO_3) is used as a carbon source [35]. In some papers, different kinds of water-soluble additives have been used to control the crystal formation of the $CaCO_3$, such as sucrose [37], anion surfactants [38], sodium dodecylbenzene sulfonate [39] and para-aminobenzoic acid [40]. Nanobubbles are clean and environmentally friendly. On the other hand, there is little research on the effect of nanobubbles on the crystal formation of calcium carbonate.

Due to the stability of the high zeta potential and the nanobubble bridging capillary force, nanobubbles could affect the crystal transition of calcium carbonate. According to previous studies [8], nanobubbles containing different gases, except CO_2, had similar physical properties. Therefore, only air nanobubbles were applied in the present study. In future studies, CO_2 and other gases would be discussed.

In this paper, air nanobubbles were produced using a hydraulic cavitation method [41,42]. The double decomposition method was used to produce calcium carbonate. To control the formation of $CaCO_3$ crystals, nanobubbles were used instead of traditional additives in this research. The DLVO (Derjaguin–Landau–Verwey–Overbeek) theory [43,44] was used to evaluate the stability of the colloidal $CaCO_3$ particles containing electrolytes in aqueous solution, while extended DLVO theory incorporating hydrophobic interaction energy [8] was used to evaluate the stability of the nanobubbles and the interaction between nanobubbles and $CaCO_3$, because of the hydrophobic surface property of nanobubbles. According to the experimental results and calculations, the possible kinetics and mechanisms of nanobubbles' influence on the transformation were proposed in detail.

2. Materials and Methods

2.1. Materials

Deionized water with a resistivity of 18.2 $M\Omega \cdot cm^{-1}$ prepared by the Classic Water Purification System from Hitech instruments Co., Ltd. (Shanghai, China) was used for all the experiments. Calcium chloride ($CaCl_2$, AR, 96.6%, Guangdong Guanghua Sci-Tech Co., Ltd., Shantou, China), and anhydrous sodium carbonate (Na_2CO_3, AR, 99.8%, Guangdong Guanghua Sci-Tech Co., Ltd., Shantou, China) were obtained from Guangdong Guanghua Sci-Tech Co., Ltd. (Shantou, China) and were of analytical purity. The pH of the aqueous solution was modified by the addition of hydrochloric acid (HCl, AR, 36%~38%, CHRON CHEMICALS, Chengdu, China) or sodium hydroxide (NaOH, AR, 96%, Guangdong Guanghua Sci-Tech Co., Ltd., Shantou, China).

To investigate the zeta potential of the vaterite and calcite, 100% of the calcite prepared in this work was used. 94% of the artificial vaterite was prepared by bubbling CO_2 into 0.05 M $CaCl_2$ aqueous solution, and ammonia (NH_4OH, 30%) was used to adjust the solution pH to 10 while the solution mixture was vigorously stirred in a beaker continuously. After 15 min, the particles were rinsed with deionized water and filtered twice, and then dried in an oven at 110 °C for 0.5 h and then at 70 °C for 12 h.

The reaction of vaterite synthesis is as follows [34]:

$$CaCl_2 + CO_2 + 2\,NH_4OH \rightarrow CaCO_3 + 2\,NH_4Cl + H_2O\ (pH = 10)$$

According to the SEM image and the XRD pattern, shown in Figure S1, spherical calcium carbonate was vaterite with high purity (94%).

2.2. Methods

2.2.1. Preparation of Nanobubble Suspension

The bulk of nanobubbles were generated by high-speed shearing using a purpose-built device (self-made equipment) which consisted of a plastic stent, an electric high-speed motor in a plastic bracket, a glass beaker, and a blade made of stainless steel as shown in Figure S2. The working principle was as follows: high-speed rotation of the blade above the motor drive creates high- and low-pressure areas and forms pressure gradients, which forms cavities in the low-pressure zones where the nanobubbles are produced [41,45].

In a typical process, the device was operated three times by adding 700 mL deionized water at 1 min intervals before producing nanobubbles. The motor frequency was initially set to 20,000 rpm. After this, 700 mL deionized water was poured into the glass beaker. The lid was put on the beaker to prevent the gas from escaping and avoid impurities in the water. The device was operated ten times, with 1 min intervals. Nanobubbles were identified after completing these steps. After generation, the solution was transferred to another glass container and sealed for the next experiment.

2.2.2. Calcium Carbonate Synthesis

The flow chart of calcium carbonate synthesis is shown in Figure 1. Firstly, nanobubbles were produced, $CaCl_2$ and Na_2CO_3 were then dissolved into the deionized water containing nanobubbles at 20 °C. The initial concentrations of $CaCl_2$ and Na_2CO_3 in deionized water were 0.05 M (M = mol/L). 200 mL Na_2CO_3 solution (0.05 M) was added to 200 mL $CaCl_2$ solution (0.05 M), and the solution mixture was vigorously stirred in the beaker continuously. Upon the completion of the designated reaction time (i.e., 0.5, 1, 2, 3, 4 h), the particles were rinsed with deionized water and filtered twice. After the last washing and suction filtration, the sedimented particles were dried in an oven at 110 °C for 0.5 h and then at 70 °C for 12 h before particle characterization.

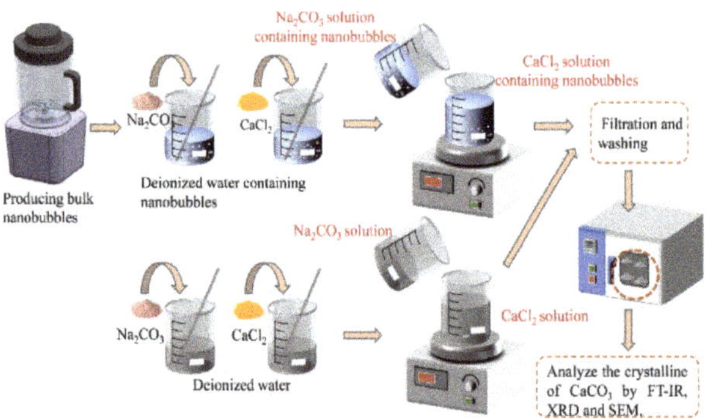

Figure 1. The flow chart of the synthesis of CaCO$_3$.

2.2.3. Characterization

The size of the nanobubbles was measured by the dynamic light scattering method (DLS, NanoBrook Omni, Brookhaven Instruments, Holtsville, NY, USA) and the concentration of nanobubbles was measured by the particle trajectory method (Nano-Sight, NS300, Malvern). The zeta potential measurements were made with the use of the phase analysis light scattering method (NanoBrook Omni, Brookhaven Instruments). After the synthesized CaCO$_3$ particles were deposited onto a stub and coated with gold, the polymorphs and morphologies were characterized by scanning electron microscopy (SEM, SEM-EDS, Phenom ProX, Netherlands) at an accelerating voltage of 15 kV. The samples obtained in Section 2.2.2 were directly used for XRD measurements. The X-ray diffraction (XRD, Rigaku D/MAX 2500V, Japan's neo-Confucianism) was used to verify the existence of vaterite and calcite and calculate the mass fraction of calcite within the synthesized CaCO$_3$ at Cu Kα target (λ = 0.1542 nm), at 40 kV and 30 mA with a scanning speed of 8° 2θ/min. The CaCO$_3$ particles with potassium bromide tableting were measured by Fourier transform infrared spectroscopy (FT-IR, Nicolet iS 50, Thermo Scientific) to verify the existence of vaterite and calcite using the transmission mode with a scanning speed of 20 spectra/s. Typical spectral resolution was 0.25 cm^{-1}.

To quantify different crystalline calcium carbonates and calculate the mass fraction of calcite within the synthesized CaCO$_3$, the intensities of the (110), (112), and (114) crystallographic planes for vaterite (I_{110V}), (I_{112V}), and (I_{114V}), and the (104) crystallographic plane for calcite (I_{104C}), the semi-quantitative phase compositions were calculated by using Equations (1) and (2) [46]:

$$X_v = (I_{110v} + I_{112v} + I_{114v})/(I_{104c} + I_{110v} + I_{110v} + I_{110v}) \tag{1}$$

$$X_c = 1 - X_v \tag{2}$$

where X_v is the mass fraction of vaterite, X_c is the mass fraction of calcite, I_{104C} is the intensities of the (104) crystallographic plane for calcite, and I_{110v}, I_{112v} and I_{114v} are the intensities of the (110), (112), and (114) crystallographic planes for vaterite, respectively.

3. Results

3.1. Generation of Air Nanobubbles

Because the whole reactor was sealed, impurities were unlikely to be found in this solution. Furthermore, the content of impurities in water was extremely low, indicating that the colloid particles observed in the solution were nanobubbles. Due to the Tyndall effect based on the light scattering of colloidal sized objects being illuminated with a laser beam after high-speed shearing, a clear light path was observed in the solution, demonstrating that plenty of colloid particles were produced. In this experiment, the hydrodynamic diameter of the air nanobubbles was measured by DLS. The mean diameter of the nanobubbles was about 83.6 nm in the absence of any salt addition. As shown in Table 1, in the 0.05 M $CaCl_2$ and 0.05 M Na_2CO_3 nanobubble aqueous solutions, the mean diameter was 126.2 nm and 101.8 nm, respectively. The pH of the Na_2CO_3 aqueous solution was 11.7, while the pH of the deionized water and $CaCl_2$ aqueous solution was about 6. As Ca^{2+} tended to be adsorb onto the surface of nanobubbles in a $CaCl_2$ aqueous solution, the nanobubbles' zeta potential was positive (7.3 mV). As shown in Figure 2, the concentration of nanobubbles in deionized water was 1.6×10^6, which was much lower than the prepared nanobubble water. According to previous studies [2,43], the mixing of organic solvents with pure water leads to the spontaneous formation of nanobubbles, while nanobubbles cannot form spontaneously in deionized water. the concentration of the bulk of the nanobubbles shifted to the right in the 0.05 M $CaCl_2$ and 0.05 M Na_2CO_3 solutions, which indicated that their mean diameters increased. Simultaneously, in the presence of a strong electrolyte, the mono-dispersity and the total concentration of the bulk nanobubbles reduced, especially noticeable in the 0.05 M $CaCl_2$ solution.

Table 1. Mean diameter, zeta potential, and concentration of nanobubbles in deionized water, in 0.05 M $CaCl_2$ aqueous solution or 0.05 M Na_2CO_3 aqueous solution.

	Mean Diameter (nm)	Zeta Potential(mV)	Concentration (bubbles/mL)	pH
In deionized water	83.6	−23.0	2.96×10^8	6.3
In 0.05 M $CaCl_2$ aqueous solution	126.2	7.3	1.48×10^8	5.6
In 0.05 M Na_2CO_3 aqueous solution	101.8	−17.8	2.49×10^8	11.7

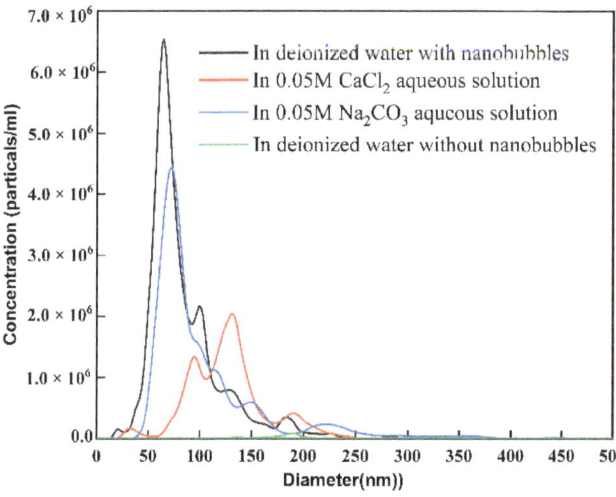

Figure 2. Size distribution of nanobubbles in deionized water with nanobubbles, in 0.05 M $CaCl_2$ aqueous solution, 0.05 M Na_2CO_3 aqueous solution or in deionized water without nanobubbles.

3.2. The Influence of Nanobubbles on the Transformation from Vaterite to Calcite

To study the influence of nanobubbles on the transformation from vaterite to calcite, deionized water or nanobubble water was used as a solvent. The products obtained with different reaction times in the presence/absence of nanobubbles were characterized by XRD. The XRD patterns of the obtained samples with different reaction times at 25 °C in nanobubble water are given in Figure 3a. In comparison with their standard JCPDS files (vaterite, 720506) and (calcite, 830577), at reaction times 0.5, 1, 2, and 3 h, a mixture of calcite and vaterite was identified. The peaks corresponding to vaterite vanished as the reaction time progressed, while the peak at $2\theta = 29.46$ corresponding to calcite became sharper and higher. After 4 h, pure calcite was identified. However, the XRD patterns of the obtained samples with different reaction times in deionized water in Figure 3b indicated that the peaks corresponding to calcite rapidly increased in intensity as vaterite transformed into calcite, but after 4 h, a mixture of calcite and vaterite was still identified.

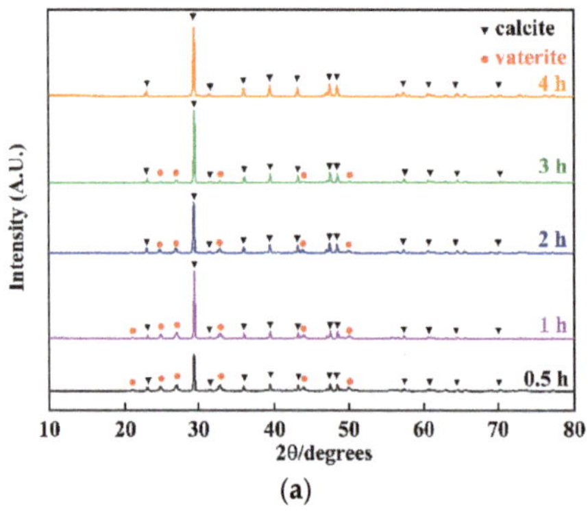

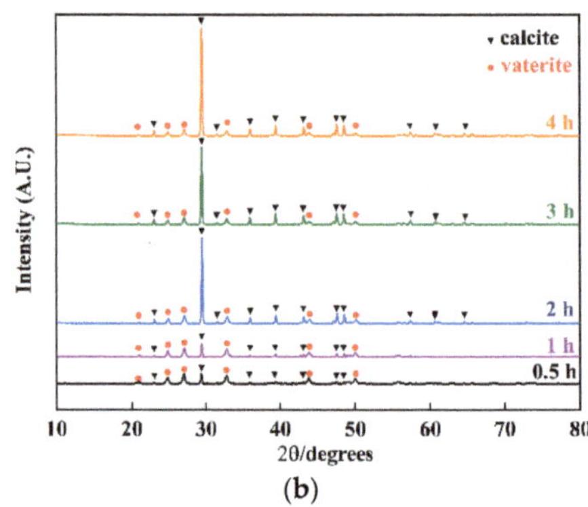

Figure 3. XRD patterns of $CaCO_3$ particles obtained with different reaction times in deionized water containing nanobubbles (**a**) and deionized water (**b**).

Comparing Figure 3a,b, when the deionized water containing nanobubbles was used as a solvent, the transformation from vaterite to calcite was completed in less time, indicating that nanobubbles could enhance this transformation.

According to the XRD patterns, the phase compositions calculated from Equations (1) and (2) are shown in Figure 4, which demonstrates the change in the mass fraction of calcite. Regardless of whether pure water or nanobubble water was used as the solvent, the vaterite mass fraction decreased and vanished with reaction time. After 1 h, the rate of transformation from vaterite to calcite decreased, and this might be because the nanobubbles gradually decreased with the progress of the reaction, and their effect on the transformation became weaker. After 4 h, when nanobubble water was used as the solvent, all the vaterite was converted to calcite. However, when deionized water was used as the solvent, a mixture of calcite and vaterite remained, with the mass of calcite accounting for only about 60% after 4 h, and took more than 6 h to finish this transformation from vaterite to calcite. The transformation from vaterite to calcite is a spontaneous process. The driving force for the acceleratory transformation by the present of nanobubbles was the vaterite–nanobubble coagulation. This is described in the following discussion Section 4.

To confirm whether vaterite remains in the products as the reaction progresses, FT-IR spectra of $CaCO_3$ were obtained after 4 h in the nanobubble water and deionized water and are plotted in Figure 5. The absorption peaks at 709 cm^{-1} indicate the presence of calcite and the peak at 748 cm^{-1} verifies the presence of vaterite [46–48]. Figure 5 indicates the presence of calcite in deionized water containing nanobubbles (a) while the presence of a mixture of calcite and vaterite in deionized water (b), and these results were in good accordance with the XRD results shown in Figure 3. The absorption peak at 709 cm^{-1} (in-plane vibration) of calcite (b) obtained in deionized water is slightly red shifted, compared with that of sample (a). The phase transformation from vaterite to calcite might have caused the peak shift due to the slight change in the atomic distance [49].

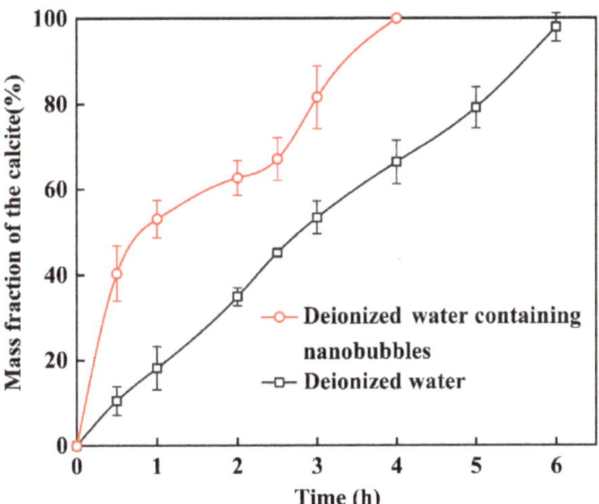

Figure 4. Mass fraction of the calcite ($W_{calcite}/W_{calcite+vaterite}$) as a function of the reaction time in deionized water containing nanobubbles and deionized water.

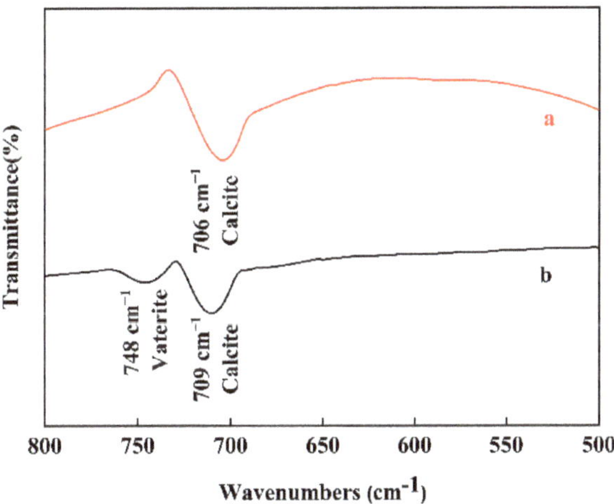

Figure 5. FT-IR spectra of $CaCO_3$ obtained after 4 h in deionized water containing nanobubbles (a) and deionized water (b).

SEM images of the fabricated samples are shown in Figure 6, with different reaction times in nanobubble water (a) to (d) and deionized water in the absence of nanobubbles (e) to (h). In deionized water containing nanobubbles, a mixture of spherical particles, i.e., crystals of vaterite [22,25,50], and cubic crystals, i.e., crystals of calcite [22,25,50], were found as shown in Figure 6a–c. It was obvious that the number of cubic crystals in deionized water containing nanobubbles increased over time with only cubic crystals present after 4 h, as shown in Figure 6d. On the other hand, in deionized water without nanobubbles, all of the products obtained at different reaction times were a mixture of spherical particles and cubic particles as shown in Figure 6e–h. The results obtained by SEM images agreed well with the results obtained by XRD (Figures 3 and 4) and FT-IR (Figure 5).

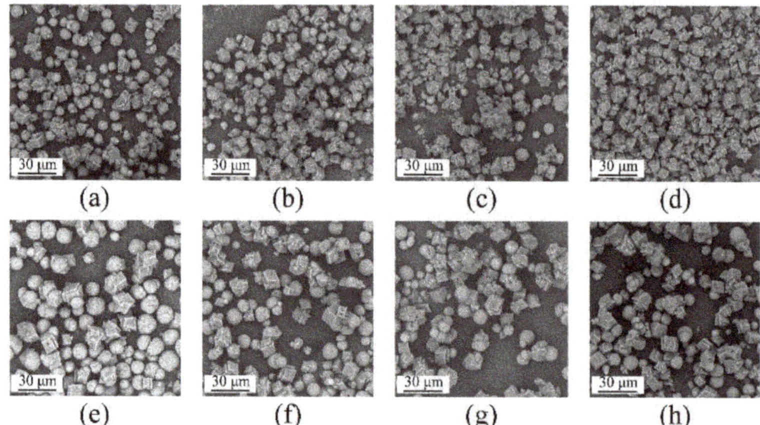

Figure 6. SEM images of CaCO$_3$ obtained in deionized water containing nanobubbles at different reaction times 1 h (**a**), 2 h (**b**), 3 h (**c**), and 4 h (**d**) and in deionized water at different reaction times 1 h (**e**), 2 h (**f**), 3 h (**g**) and 4 h (**h**).

3.3. The Influence of Concentration of Nanobubbles on the Transformation from Vaterite to Calcite

To study the effect of nanobubbles on the polymorphs of CaCO$_3$, deionized water with different concentrations of nanobubbles was used as a solvent. In this experiment, high-speed shearing was used to produce nanobubbles. The same motor frequency and shearing time were chosen to avoid multiple variables occurring at the same time.

The volume ratio of deionized water to nanobubbles in groups 1 to 4 is listed in Table 2. The low concentration of nanobubbles in deionized water (Groups 2 and 3) was prepared by adding deionized water containing nanobubbles obtained by high-speed shearing in different volumes to the deionized water. In groups 2 and 3, the volume ratio of pure water to nanobubbles (V$_{\text{deionized water containing nanobubble}}$: V$_{\text{deionized water}}$) was 1:3 and 1:1, respectively. In group 1, only deionized water was used as a solvent, while in group 4, nanobubble water without dilution was used. The total nanobubble number was from 0 to 2.96×10^8 bubbles/mL.

Table 2. The concentration of nanobubbles in deionized water in groups 1 to 4.

	The Volume Ratio of Pure Water to Nanobubbles (V$_{\text{deionized water containing nanobubble}}$: V$_{\text{deionized water}}$)	The Concentration of Nanobubbles (Bubbles/mL)
Group1	Deionized water	0
Group2	1:3	4.59×10^7
Group3	1:1	1.29×10^8
Group4	Initial deionized water containing nanobubbles	2.96×10^8

The XRD patterns of the obtained samples with different concentrations of nanobubbles in deionized water are shown in Figure 7. When the concentration of nanobubbles (groups 1 to 3) were low, mixtures composed of vaterite and calcite were obtained. When the concentration of nanobubbles was 2.96×10^8 bubbles/mL (group 4), only peaks corresponding to calcite were found.

According to the calculations using Equations (1) and (2), the mass fraction of the calcite (W$_{\text{calcite}}$/W$_{\text{calcite+vaterite}}$) as a function of the concentration of nanobubbles in deionized water after 4 h is shown in Figure 8. As the concentration of nanobubbles increased, the mass fraction of calcite increased linearly and reached 100% when the concentration of nanobubbles was at 2.96×10^8 bubbles/mL.

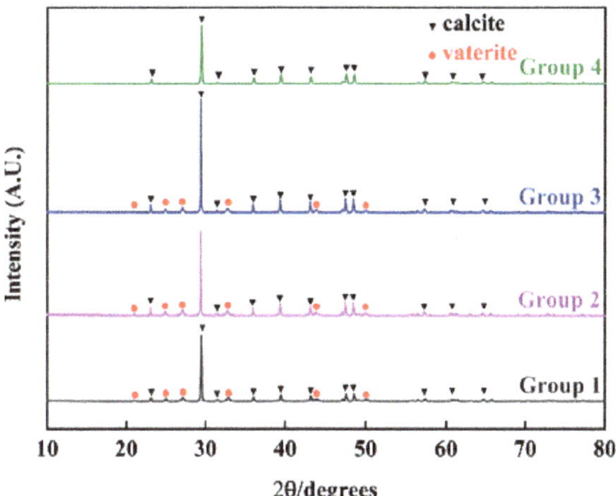

Figure 7. XRD patterns of $CaCO_3$ particles obtained after 4 h as a function of the concentration of nanobubbles in deionized water in groups 1 to 4.

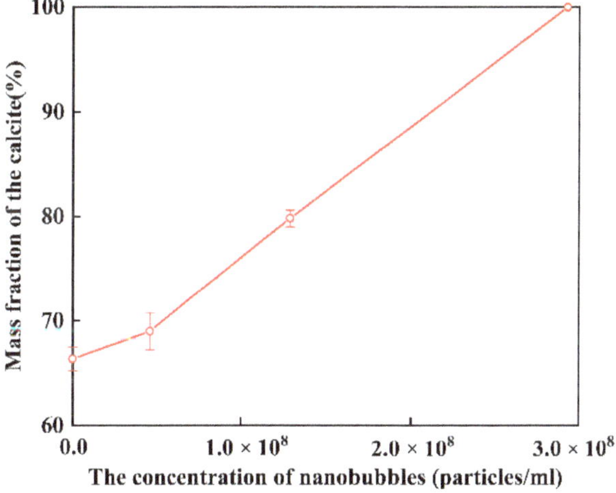

Figure 8. Mass fraction of the calcite ($W_{calcite}/W_{calcite+vaterite}$) as a function of the concentration of nanobubbles in deionized water after 4 h.

SEM images of $CaCO_3$ obtained after 4 h in groups 1–4 are shown in Figure 9a–d, respectively. In Figure 9a–c, the products obtained were a mixture of spherical vaterite particles and cubic calcite particles, while in deionized water containing a high concentration of nanobubbles (2.96×10^8 bubbles/mL), only cubic calcite particles were observed, as shown in Figure 8. The results obtained by SEM images were quite consistent with those obtained by XRD, as shown in Figure 7. The size of calcite and vaterite particles was almost comparable in deionized water containing various concentrations of nanobubbles. The size distribution of calcite after 4 h prepared in 2.96×10^8 bubbles/mL of nanobubble water is shown in Figure S2a. The mean diameter of calcite was around 5 μm.

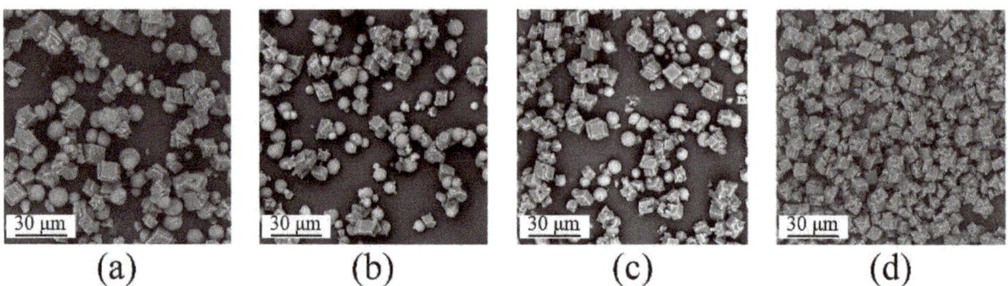

Figure 9. Mass fraction of the calcite ($W_{calcite}/W_{calcite+vaterite}$) as a function of the concentration of nanobubbles in deionized water after 4 h, group 1 (**a**), group 2 (**b**), group 3 (**c**), group 4 (**d**).

4. Discussion

4.1. CaCO$_3$ Particle Synthesis with Different Nanobubble Concentrations

The reaction of CaCO$_3$ synthesis is as follows:

$$Na_2CO_3 + CaCl_2 \rightarrow CaCO_3 + 2NaCl \tag{3}$$

During this reaction, 0.05 M of Na$_2$CO$_3$ aqueous solution was reacted with an equal amount of CaCl$_2$ aqueous solution of the same concentration, the synthesized CaCO$_3$ was 0.025 M (C_1). One particle diameter (d) of synthesized CaCO$_3$ particle was around 5 μm as shown in Figure S2a. The average density of CaCO$_3$ (ρ) was about 2.7 g/mL (calcite 2.71, vaterite 2.65 [51]). The total number of CaCO$_3$ particles synthesized (N_{CaCO3}) is as follows,

$$N_{CaCO_3} = V/V_{CaCO_3} = 6M_{CaCO_3} \times C_1 \times V_{solution}/(\pi \rho d^3) \tag{4}$$

The concentration of CaCO$_3$ particles (C, particles/mL) is given by Equation (5),

$$C = N_{CaCO_3}/V_{solution} = 6M_{CaCO_3} \times C_1/(\pi \rho d^3) = 1.4 \times 10^7 \text{particles/mL} \tag{5}$$

where V is the total volume of calcium carbonate obtained, V_{CaCO_3} is the volume of a single calcium carbonate particle, M_{CaCO_3} is the molecular weight of calcium carbonate, $V_{solution}$ is the volume of the solution, and d is the mean diameter of the calcium carbonate particles.

As the nanobubble number was 2 to 3 × 10^8 bubbles/mL as shown in Table 2, the number of nanobubbles applied for this experiment was 10 times as many as that of CaCO$_3$ particles.

The interaction between nanobubbles and CaCO$_3$ particles is important for the preparation of calcite and vaterite. After the reaction shown in Equation (3), the pH equilibrated around 10. Figure 10 shows the zeta potential of air nanobubbles, calcite, and vaterite in the 0.05 M of NaCl aqueous solution containing different concentrations of Ca^{2+} at pH = 10. At pH = 10, the zeta potential of vaterite was always positive, while calcite and air nanobubbles were negative at concentrations less than 10^{-3} M of Ca^{2+}.

After the reaction described in Equation (3), the Ca^{2+} concentration in the solution was constant at 2.5 × 10^{-5} M after 1.5 h, as shown in Figure S4. Therefore, after 1.5 h the zeta potential of calcite and nanobubbles was negative and the zeta potential of vaterite was positive, as shown in Figure 10.

4.2. The Interaction between Nanobubbles, CaCO$_3$ Particles and Nanobubble-CaCO$_3$ Particles

In this section, we evaluate (a) bubble–bubble interaction, (b) vaterite–calcite interaction, (c) bubble–calcite interaction, and (d) bubble–vaterite interaction, by using the classical and extended DLVO theory in order to investigate and propose the mechanism of the transformation from vaterite to calcite in the presence of air nanobubbles.

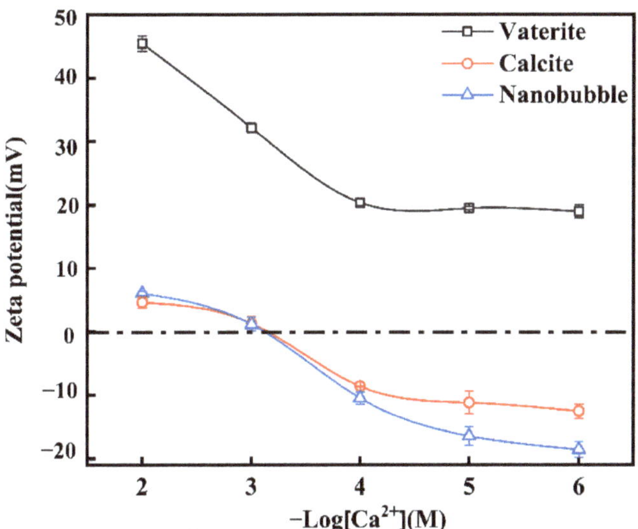

Figure 10. Zeta potential of air nanobubble, calcite, and vaterite in a 0.05 M of NaCl aqueous solution containing various concentrations of Ca^{2+} at pH = 10.

In addition to the classical DLVO theory, i.e., the summation of the attractive van der Waals interaction energy (E_A) and the repulsive electrostatic interaction energy (E_R), to evaluate the interactions between different $CaCO_3$ particles (vaterite-calcite), we used the hydrophobic interaction energy (E_h) to evaluate the stability of the nanobubbles. The total potential energy (E_T) between two nanobubbles is described as follows [8,52,53]:

$$E_T = E_A + E_R + E_h \tag{6}$$

$$\frac{E_T}{kT} = \frac{E_A + E_R + E_h}{kT} \tag{7}$$

The total potential energy (E_T) is a key indicator of the stability of nanobubbles and can evaluate whether the nanobubbles coagulate or disperse. When E_T is significantly positive (e.g., >15 kT), there is a potential barrier between the two adjacent bubbles which will not coagulate, while if E_T is slightly positive or negative (<15 kT), the nanobubbles can coagulate together and form a larger bubble.

$E_A + E_h$ and E_R can be described by the following formulas:

$$E_A + E_h = -\frac{A+K}{6}\left\{\frac{2a_1a_2}{R^2 - (a_1 + a_2)^2} + \frac{2a_1a_2}{R^2 - (a_1 - a_2)^2} + \ln\left[R^2 - (a_1 + a_2)^2\right] - \ln\left[R^2 - (a_1 - a_2)^2\right]\right\} \tag{8}$$

$$R = a_1 + a_2 + h \tag{9}$$

$$E_R = \frac{\pi\varepsilon_r\varepsilon_0 a_1 a_2(\psi_1^2 + \psi_2^2)}{a_1 + a_2}\left\{\frac{2\psi_1\psi_2}{\psi_1^2 + \psi_2^2}\ln\frac{1 + \exp(-\kappa h)}{1 - \exp(-\kappa h)} + \ln[1 + \exp(-2\kappa h)]\right\} \tag{10}$$

where A is the Hamaker constant, K is the hydrophobic force constant, a_1 and a_2 are the radii of the nanobubbles, h is the surface distance of the nanobubbles, κ represents the Debye–Hückel parameter, and ψ_1 and ψ_2 represent the surface potential of the nanobubbles of radii a_1 and a_2, respectively. The thickness of the electric double layer can be represented by the Debye length ($\lambda_D = 1/\kappa$), which can be described by the following formula:

$$\kappa = \sqrt{\frac{2nz^2e^2}{\varepsilon_r\varepsilon_0 kT}} \tag{11}$$

$$n = 1000N_A C \tag{12}$$

where n is the concentration of anions or cations in the solution, z represents the valence of an ion, e represents the electron charge, ε_r represents the relative dielectric constant, ε_0 represents the permittivity of a vacuum, T represents the temperature (K), N_A represents the Avogadro number and C represents concentration of anions or cations (mol/L).

Bubble–bubble interaction: according to our previous study [8], these nanobubbles could stay stable for several weeks or even months. Even in the presence of salt, nanobubbles can also be stable for days. They were stable enough to perform zeta potential measurements. The zeta potential of nanobubbles is shown in Figure 10. When the nanobubbles existed in the reacted solution with a Ca^{2+} concentration less than 10^{-3} M, the zeta potential of the nanobubble was negative, as shown in Figure 11. The mean diameter of the nanobubbles in the mixed vaterite and calcite suspension after reaction depending on the time measured by DLS after filtration through 10–15 μm filter paper is shown in Figure 11. The mean diameter of the nanobubbles increased from an initial 100–130 nm (Table 1) to 600–700 nm during 3 h of the reaction time (Figure 11). Figure 12a shows the total potential energy (E_T) between the bubbles calculated by using the extended DLVO theory, and there was no potential barrier to prevent the coalescence of the bubbles, which was consistent with the bubble size enlargement as shown in Figure 11.

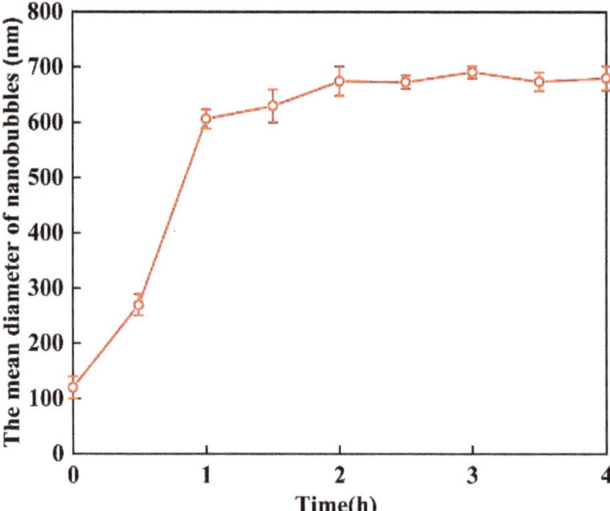

Figure 11. The mean diameter of nanobubbles in mixed vaterite and calcite suspension after reaction depending on the time measured by DLS after filtration with 10–15 μm filter paper.

Vaterite–calcite interaction: Figure 12b shows the total potential energy (E_T) between the calcite and vaterite particles (6 μm mean diameter as shown in Figure S3c) calculated by using the classical DLVO theory as this was an interaction between solid particles. There was no potential barrier when considering the low zeta potential of calcite (−10 mV) and vaterite (20 mV) indicating their coagulation, (Figure 10).

Air bubble–calcite interaction: Figure 13 shows the total potential energy (E_T) between the different sizes of air bubbles and calcite particles as a function of their distance by the extended DLVO theory. Figure 13a,b depict the interactions between the initial 120 nm and 700 nm nanobubbles after 1 h (shown in Figure 11) and the prepared 5 μm calcite after 1 h (shown in Figure S2b), respectively. In the both cases, there was no potential barrier between nanobubbles and calcite, indicating their hetero-coagulation.

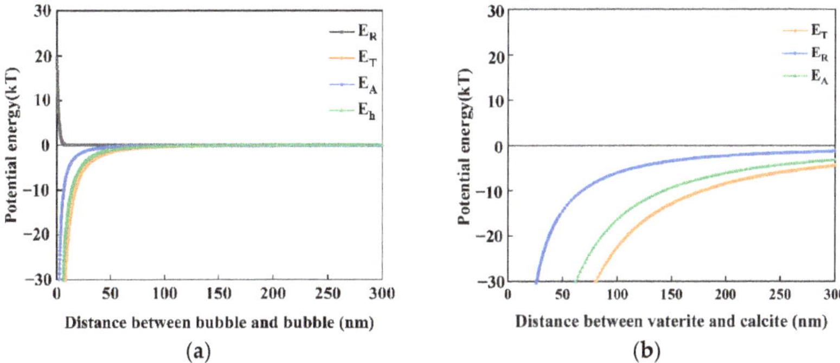

Figure 12. (a) Total potential energy E_T between bubbles (700 nm, -15 mV) as a function of the distance between two bubbles by the extended DLVO theory, at a Na^+ concertation of 0.05 M and a Ca^{2+} concentration of 2×10^{-5} M; (b) Total potential energy E_T between calcite (3 μm, -10 mV) and vaterite (6 μm, 20 mV) as a function of the distance between two particles by the classical DLVO theory, at a Ca^{2+} concentration of 2.5×10^{-5} M.

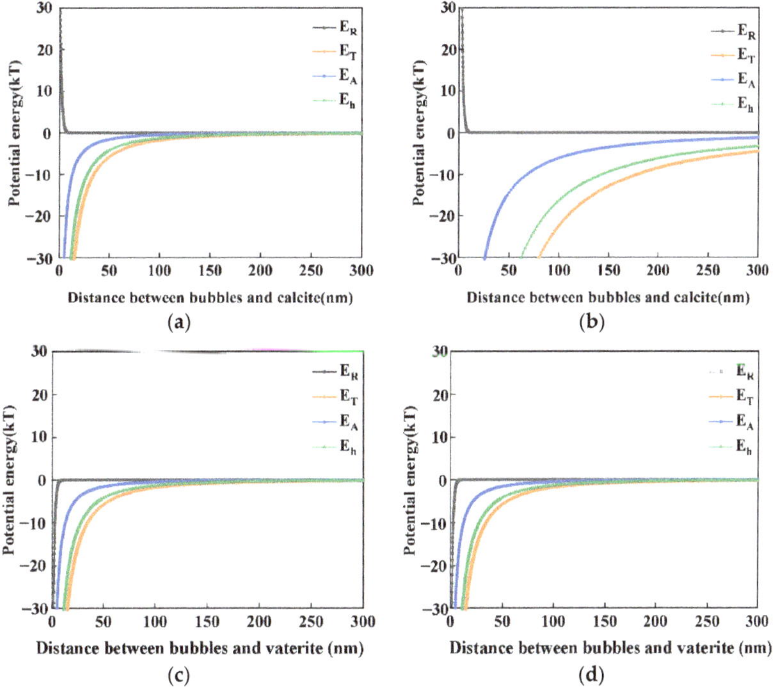

Figure 13. Total potential energy E_T between the different sizes of air bubbles and calcium carbonate particles as a function of the distance between a nanobubble and a calcium carbonate particle by the extended DLVO theory ($K = 10^{-19}$ J). Na^+ concentration was 0.05 M and Ca^{2+} concentration was 2×10^{-5} M. (a) Air nanobubbles (120 nm, -15 mV) and calcite particles (5 μm, -10 mV); (b) Air nanobubbles (700 nm, -15 mV) and calcite particle (5 μm, -10 mV); (c) Air nanobubbles (120 nm, -15 mV) and vaterite particles (6 μm, 20 mV); (d) Air nanobubbles (700 nm, -15 mV) and vaterite particles (6 μm, 20 mV).

Air bubble–vaterite interaction: considering the zeta potential of the vaterite particles was positive (20 mV) and air bubbles was negative, as shown in Figure 13c,d, there was no potential barrier between the nanobubbles and vaterite indicating hetero-coagulation between nanobubbles and vaterite.

To sum up, in deionized water containing nanobubbles, according to the calculations based on the extended DLVO theory, there was no potential barrier between nanobubbles, calcite or vaterite, indicating that negatively charged nanobubbles could attach and accumulate not only on the surfaces of calcite particles but also on the surfaces of vaterite particles. This supported the premise that nanobubbles could bind calcite with vaterite.

4.3. The Capillary Force Model between Calcite and Vaterite

Figure 14 shows SEM images of spherical vaterite particles attached to the cubic surface of calcite particles in deionized water (a) and nanobubble-containing solution (b) after 0.5 h. Obviously, in the nanobubble-containing solution (b), there were more spherical vaterite particles attached to the surface of cubic calcite particles than that in the deionized water without nanobubbles (a). The zeta potential of the vaterite particles was positive, while the surface of the calcite particles was negative. The vaterite particles tended to be clustered with calcite particles. Furthermore, compared to vaterite particles, calcite particles were cubic crystals, which had a higher probability of coming into contact with vaterite particles. This indicates that nanobubbles can bind calcite with vaterite, enabling vaterite and calcite to clustered together in deionized water containing nanobubbles.

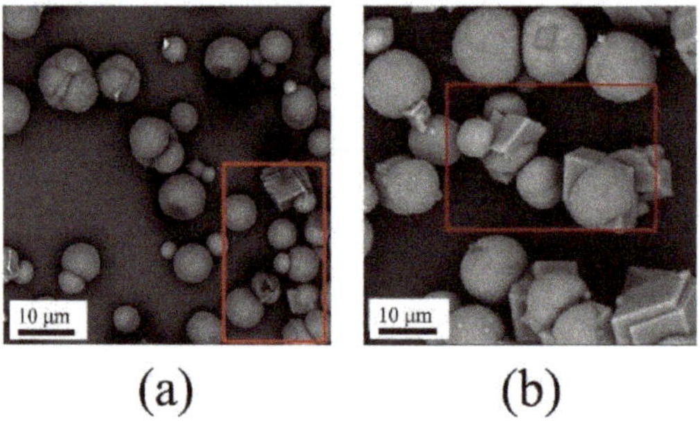

Figure 14. SEM images of spherical vaterite particles attached onto cubic calcite particle surfaces in deionized water (**a**) and in deionized water containing nanobubbles (**b**) (after 0.5 h passed).

The capillary force model [54–57], based on the forces caused by the pressure drop across the interface at the neck and the interfacial tension force, was used to calculate the capillary force between the vaterite and calcite particles:

$$F_c = \pi\sigma \sin \alpha [2\sin(\alpha + \theta) + R\sin\alpha(1/r - 1/l)] \tag{13}$$

$$r = [R(1 - \cos \alpha) + h]/[\cos(\alpha + \theta) + \cos \alpha] \tag{14}$$

$$l = R\sin\alpha - r[1 - \sin(\alpha + \theta)] \tag{15}$$

where F_c is the capillary force, σ is the surface tension coefficient of the liquid, α is the fill angle, R is the radius of the colloidal probe, θ is the liquid contact angle, r and l are the principal radii of the capillary bridge, and h is the separation distance between the vaterite and calcite surfaces, as shown in Figure 15a.

In absence of nanobubbles, spherical vaterite attached to the surface of cubic calcite might become separated by the repeated collisions between solid particles and high-speed flowing solution stirred under the magnetic agitator [58]. It can be difficult for ultrafine particles to collide and coagulated. On the other hand, in presence of nanobubbles, the negatively charged nanobubbles can attach and accumulate on the surface of calcite particles and this aggregate can further coagulate with positively charged vaterite particles [58].

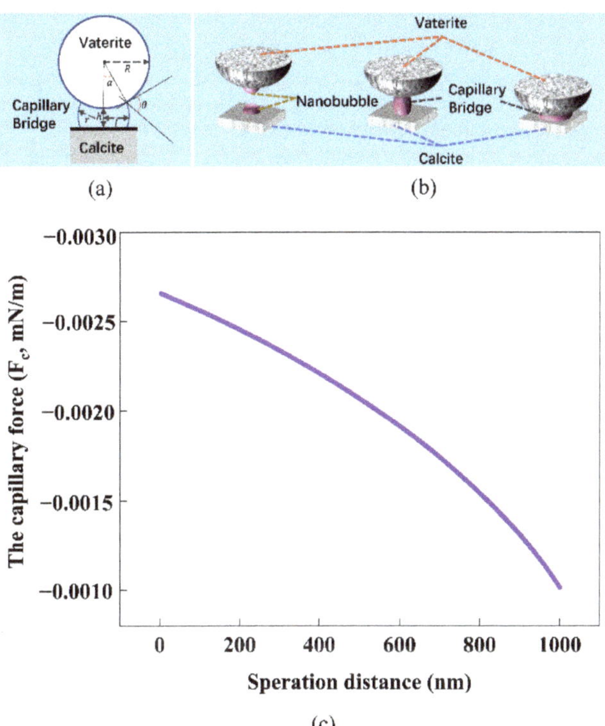

Figure 15. The geometry of vaterite and calcite particles used to calculate the capillary force between them (**a**); simplified mechanism of a nanobubble bridging the surfaces of vaterite and calcite particles (**b**); the nanobubble capillary force bridging vaterite and calcite particles as a function of their surface distance (**c**).

According to previous studies, the probability of collision between a particle and a bubble is closely related to bubble size [59]. Because of the smaller contact area and aggregation effects on ultrafine particles, nanobubbles are more easily attached to the surface of vaterite and calcite particles [13] in comparison with larger bubbles. According to the nanobubble bridging capillary force (NBCF) resulting from the long range hydrophobic attractive force [55,60–62], as they approached, a gaseous capillary bridge could be formed through the coalescence of nanobubbles. The underlying mechanism is described in Figure 15b.

The induced attractive capillary force was helpful to enhance the aggregation of vaterite and calcite particles through bubble bridging. The nanobubble bridging capillary force as a function of the surface separation distance between vaterite and calcite particles is shown in Figure 15c. Even if the separation distance was 1000 nm, F_c was still 0.001 mN/m, which would avoid separation of concentrated solid particles and enhance the effective slip lengths between solid and liquid, enhancing the aggregation between solid particles [63]. Once vaterite and calcite particles aggregated together, the capillary forces produced in the presence of the nanobubbles prevented them from separating. The calculation results agreed well with the results obtained by SEM images (Figure 14). There were more spherical vaterite particles attached to the surfaces of cubic calcite particles in deionized water in the presence of nanobubbles than that in deionized water in the absence of nanobubbles.

4.4. Formation Mechanism of the Crystallization Transformation of CaCO$_3$

In this series of experiments reported in this article, the double decomposition method was used to produce calcium carbonate. After mixing CaCl$_2$ and Na$_2$CO$_3$ solution, amorphous calcium carbonate (ACC) formed in several seconds, then ACC transformed into vaterite, and finally, all of them changed into calcite. This transformation occurred in two stages: the first step was the transformation from ACC to vaterite, and the second step was the rate-determining process from vaterite to calcite. Gas–water interfaces might act as nucleation sites, which would enhance the transformation from ACC to vaterite. According to the literature [22,23,64], the second stage of the reaction is much slower than the first stage. However, the second step directly determines the formation of calcite. Through the dissolution–reprecipitation mechanism, vaterite can be transformed into calcite [22,23,64]. For detail, the process was divided into two steps, the first step was the dissolution of vaterite, and the second step was the growth of the calcite, which was the rate-determining process. The second step was the recrystallization of calcium carbonate which was divided into the dissolution of vaterite and the growth of calcite [22,50,64]. As shown in Figure 16, Vaterite can dissolve in water and release calcium and carbonate ions, which are then reprecipitated on the surfaces of the growing calcite crystals. As shown in Figures 4 and 8, after 4 h, all the vaterite was converted to calcite in deionized water containing nanobubbles, while, in absence of nanobubbles, a mixture of calcite (60%) and vaterite (40%) remained. With the concentration of nanobubbles increasing, the mass fraction of calcite increased linearly and reached 100% when the concentration of nanobubbles was at 2.96×10^8 bubbles/mL. The results show that nanobubbles can enhance the transformation from vaterite to calcite. According to the calculation shown in Figure 13, there was no potential barrier between nanobubble, calcite and vaterite, indicating heterocoagulation between them. The number of nanobubbles applied for this experiment was 10 times as many as that of CaCO$_3$ particles. Under the intense stirring of the magnetic agitator, nanobubbles improved the probability of collision. Due to the nanobubble bridging capillary force [63] increases with shorter separation distances between the vaterite and calcite particles, as shown in Figure 15c; thus, their aggregation can become more stable. As clearly shown in Figure 14, it was easier and more stable for vaterite and calcite to become aggregated together in the presence of nanobubbles (a) compared with the absence of nanobubbles (b). This aggregation allowed more Ca^{2+} to attach to the surface of calcite, allowing calcite to grow more efficiently and transform faster.

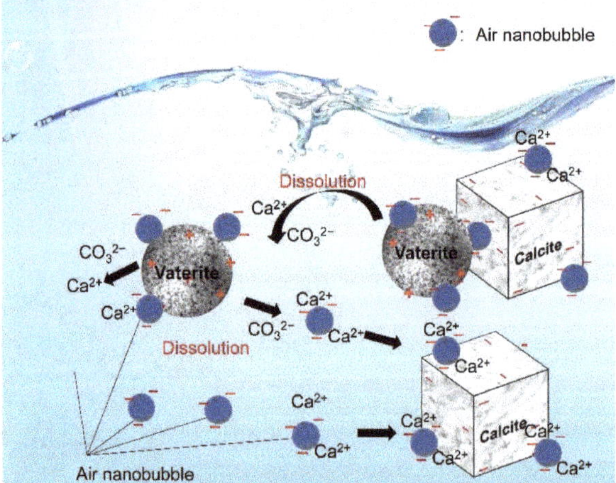

Figure 16. The interaction between nanobubble, vaterite, and calcite and transformation from vaterite to calcite. (Ca^{2+} concentration 10^{-4} to 10^{-6} M, pH = 10, vaterite mean diameter 5 μm, air nanobubble 120 nm.).

5. Conclusions

This was the first study using nanobubbles to aid the transformation of vaterite to calcite, and we confirmed that the transformation was accelerated in the presence of air nanobubbles. According to the calculations based on the extended DLVO theory, in the reacted solution, there was no potential barrier between the nanobubbles and both calcite or vaterite particles, indicating their coagulation. In deionized water without nanobubbles, spherical vaterite attached on the surface of cubic calcite particles of about 5 μm in size but would be dispersed by the repeated collisions between solid particles and high-speed flowing solution. However, according to the calculations of the nanobubble bridging capillary force and the SEM images of synthesized calcite particles, nanobubbles played an important role as the binder in strengthening the coagulation between calcite and vaterite. Therefore, the transformation rate increased in deionized water containing nanobubbles, compared with the aqueous solution without nanobubbles. Based on the results, this research provides a novel method for controlling the crystal morphology and reaction kinetics of $CaCO_3$. The CO_2 bubbling method (indirect carbonation) is an important way to synthesize vaterite. The size of the bubbles has an obvious effect on the vaterite content and can be controlled by adjusting the ventilation equipment. In future studies, the influence of the bubble size on crystal formation of calcium carbonate should be studied to produce pure vaterite.

Supplementary Materials: The following supporting information can be downloaded at: https://www.mdpi.com/article/10.3390/ma15217437/s1, Figure S1: SEM images and XRD pattern of vaterite obtained; Figure S2: The schematic diagram of the machine for producing bulk nanobubbles (a), and the size of the stainless-steel blade and beaker (b); Figure S3: Size distribution of calcite after 4 h (a), the size distribution of calcite after 1 h, and size distribution of vaterite after 1 h prepared in 2.96×10^8 particles/mL of nanobubble water; Figure S4: The concentration of Ca^{2+} depending on the time in the solution filtrated at 0.45 μm after the reaction to prepare $CaCO_3$;

Author Contributions: Y.W.: Data curation, Investigation, Writing original draft. M.H.: Investigation. C.H.: Review and editing. K.W.: Review and editing. N.T.H.N.: Writing—Review and editing. S.L.: Review and editing. G.D.: Review and editing. A.O.: Review and editing. T.F.: Writing—Review and editing. All authors have read and agreed to the published version of the manuscript.

Funding: This research was funded by the Natural Science Foundation of China, grant number NSFC 21976039.

Institutional Review Board Statement: Not applicable.

Informed Consent Statement: Not applicable.

Data Availability Statement: The authors confirm that the data supporting the findings of this study are available within the article.

Acknowledgments: We appreciate the special funding support by "Guangxi Bagui scholars".

Conflicts of Interest: The authors declare no conflict of interest.

References

1. Gurung, A.; Dahl, O.; Jansson, K. The fundamental phenomena of nanobubbles and their behavior in wastewater treatment technologies. *Geosystem. Eng.* **2016**, *19*, 133–142. [CrossRef]
2. Michailidi, E.D.; Bomis, G.; Varoutoglou, A.; Kyzas, G.Z.; Mitrikas, G.; Mitropoulos, A.C.; Efthimiadou, E.K.; Favvas, E.P. Bulk nanobubbles: Production and investigation of their formation/stability mechanism. *J. Colloid. Interf. Sci.* **2020**, *564*, 371–380. [CrossRef] [PubMed]
3. Nirmalkar, N.; Pacek, A.W.; Barigou, M. On the Existence and Stability of Bulk Nanobubbles. *Langmuir* **2018**, *34*, 10964–10973. [CrossRef] [PubMed]
4. Ulatowski, K.; Sobieszuk, P.; Mróz, A.; Ciach, T. Stability of nanobubbles generated in water using porous membrane system. *Chem. Eng. Process.* **2019**, *136*, 62–71. [CrossRef]
5. Ducker, W.A. Contact angle and stability of interfacial nanobubbles. *Langmuir* **2009**, *25*, 8907–8910. [CrossRef]

6. Ma, X.; Li, M.; Pfeiffer, P.; Eisener, J.; Ohl, C.D.; Sun, C. Ion adsorption stabilizes bulk nanobubbles. *J. Colloid. Interf. Sci.* **2022**, *606 Pt 2*, 1380–1394. [CrossRef]

7. Zhang, H.; Guo, Z.; Zhang, X. Surface enrichment of ions leads to the stability of bulk nanobubbles. *Soft. Matter.* **2020**, *16*, 5470–5477. [CrossRef]

8. Zhou, Y.; Han, Z.; He, C.; Feng, Q.; Wang, K.; Wang, Y.; Luo, N.; Dodbiba, G.; Wei, Y.; Otsuki, A.; et al. Long-Term Stability of Different Kinds of Gas Nanobubbles in Deionized and Salt Water. *Materials* **2021**, *14*, 1808. [CrossRef] [PubMed]

9. Agarwal, A.; Ng, W.J.; Liu, Y. Principle and applications of microbubble and nanobubble technology for water treatment. *Chemosphere* **2011**, *84*, 1175–1180. [CrossRef]

10. Mai, L.; Yao, A.; Li, J.; Wei, Q.; Yuchi, M.; He, X.; Ding, M.; Zhou, Q. Cyanine 5.5 conjugated nanobubbles as a tumor selective contrast agent for dual ultrasound-fluorescence imaging in a mouse model. *PLoS ONE* **2013**, *8*, e61224. [CrossRef] [PubMed]

11. Cavalli, R.; Bisazza, A.; Giustetto, P.; Civra, A.; Lembo, D.; Trotta, G.; Guiot, C.; Trotta, M. Preparation and characterization of dextran nanobubbles for oxygen delivery. *Int. J. Pharm.* **2009**, *381*, 160–165. [CrossRef]

12. Batagoda, J.H.; Hewage, S.A.; Meegoda, J.N. Remediation of heavy-metal-contaminated sediments in USA using ultrasound and ozone nanobubbles. *J. Environ. Eng. Sci.* **2019**, *14*, 130–138. [CrossRef]

13. Wang, Y.; Pan, Z.; Luo, X.; Qin, W.; Jiao, F. Effect of nanobubbles on adsorption of sodium oleate on calcite surface. *Miner. Eng.* **2019**, *133*, 127–137. [CrossRef]

14. Calgaroto, S.; Wilberg, K.Q.; Rubio, J. On the nanobubbles interfacial properties and future applications in flotation. *Miner. Eng.* **2014**, *60*, 33–40. [CrossRef]

15. Khoshroo, M.; Javid, A.; Katebi, A. Effects of micro-nano bubble water and binary mineral admixtures on the mechanical and durability properties of concrete. *Constr. Build. Mater.* **2018**, *164*, 371–385. [CrossRef]

16. Bull, D.S.; Kienle, D.F.; Sosa, A.F.C.; Nelson, N.; Roy, S.; Cha, J.N.; Schwartz, D.K.; Kaar, J.L.; Goodwin, A.P. Surface-Templated Nanobubbles Protect Proteins from Surface-Mediated Denaturation. *J. Phys. Chem. Lett.* **2019**, *10*, 2641–2647. [CrossRef] [PubMed]

17. Hernandez, C.; Abenojar, E.C.; Hadley, J.; de Leon, A.C.; Coyne, R.; Perera, R.; Gopalakrishnan, R.; Basilion, J.P.; Kolios, M.C.; Exner, A.A. Sink or float? Characterization of shell-stabilized bulk nanobubbles using a resonant mass measurement technique. *Nanoscale* **2019**, *11*, 851–855. [CrossRef]

18. Kurokawa, H.; Matsui, H.; Ito, H.; Taninaka, A.; Shigekawa, H.; Dodbiba, G.; Wei, Y.; Fujita, T. Antioxidant Effect of Hydrogen Nanobubble Contributes to Suppression of Tumor Cell Growth. *Biomed. J. Sci. Tech. Res.* **2019**, *19*, 14592–14594. [CrossRef]

19. Etchepare, R.; Azevedo, A.; Calgaroto, S.; Rubio, J. Removal of ferric hydroxide by flotation with micro and nanobubbles. *Sep. Purif. Technol.* **2017**, *184*, 347–353. [CrossRef]

20. Temesgen, T.; Bui, T.; Han, M.; Kim, T.; Park, H. Micro and nanobubble technologies as a new horizon for water-treatment techniques: A review. *Adv. Colloid. Interf. Sci.* **2017**, *246*, 40–51. [CrossRef]

21. Calgaroto, S.; Azevedo, A.; Rubio, J. Separation of amine-insoluble species by flotation with nano and microbubbles. *Miner. Eng.* **2016**, *89*, 24–29. [CrossRef]

22. Rodriguez-Blanco, J.D.; Shaw, S.; Benning, L.G. The kinetics and mechanisms of amorphous calcium carbonate (ACC) crystallization to calcite, via vaterite. *Nanoscale* **2011**, *3*, 265–271. [CrossRef] [PubMed]

23. Sawada, K. The mechanisms of crystallization and transformation of calcium carbonates. *Pure. Appl. Chem.* **1997**, *69*, 921–928. [CrossRef]

24. Kim, G.; Kim, S.; Kim, M. Effect of sucrose on CO_2 storage, vaterite content, and $CaCO_3$ particle size in indirect carbonation using seawater. *J. CO_2 Util.* **2022**, *57*, 101894. [CrossRef]

25. Beck, R.; Andreassen, J. The onset of spherulitic growth in crystallization of calcium carbonate. *J. Cryst. Growth* **2010**, *312*, 2226–2238. [CrossRef]

26. Wang, W.; Wang, G.; Liu, Y.; Zheng, C.; Zhan, Y. Synthesis and characterization of aragonite whiskers by a novel and simple route. *J. Mater. Chem.* **2001**, *11*, 1752–1754. [CrossRef]

27. Han, J.; Jung, S.; Kang, D.; Seo, Y. Development of Flexible Calcium Carbonate for Papermaking Filler. *ACS Sustain. Chem. Eng.* **2020**, *8*, 8994–9001. [CrossRef]

28. Lizandara-Pueyo, C.; Fan, X.; Ayats, C.; Pericàs, M.A. Calcium carbonate as heterogeneous support for recyclable organocatalysts. *J. Catal.* **2021**, *393*, 107–115. [CrossRef]

29. Dou, J.; Zhao, F.; Fan, W.; Chen, Z.; Guo, X. Preparation of non-spherical vaterite $CaCO_3$ particles by flash nano precipitation technique for targeted and extended drug delivery. *J. Drug. Deliv. Sci. Tec.* **2020**, *57*, 101768. [CrossRef]

30. Lin, P.; Wu, H.; Hsieh, S.; Li, J.; Dong, C.; Chen, C.; Hsieh, S. Preparation of vaterite calcium carbonate granules from discarded oyster shells as an adsorbent for heavy metal ions removal. *Chemosphere* **2020**, *254*, 126903. [CrossRef]

31. Nakamura, J.; Kasuga, T.; Sakka, Y. Preparation of carbamate-containing vaterite particles for strontium removal in wastewater treatment. *J. Asian Ceram. Soc.* **2018**, *5*, 364–369. [CrossRef]

32. Song, X.; Cao, Y.; Bu, X.; Luo, X. Porous vaterite and cubic calcite aggregated calcium carbonate obtained from steamed ammonia liquid waste for Cu^{2+} heavy metal ions removal by adsorption process. *Appl. Surf. Sci.* **2021**, *536*, 147958. [CrossRef]

33. Sasamoto, R.; Kanda, Y.; Yamanaka, S. Difference in cadmium chemisorption on calcite and vaterite porous particles. *Chemosphere* **2022**, *297*, 134057. [CrossRef] [PubMed]

34. Hadiko, G.; Han, Y.; Fuji, M.; Takahashi, M. Synthesis of hollow calcium carbonate particles by the bubble templating method. *Mater. Lett.* **2005**, *59*, 2519–2522. [CrossRef]

35. Sha, F.; Zhu, N.; Bai, Y.; Li, Q.; Guo, B.; Zhao, T.; Zhang, F.; Zhang, J. Controllable Synthesis of Various CaCO$_3$ Morphologies Based on a CCUS Idea. *Chem. Eng.* **2016**, *4*, 3032–3044. [CrossRef]

36. Yang, L.; Chu, D.; Sun, H.; Ge, G. Room temperature synthesis of flower-like CaCO$_3$ architectures. *New. J. Chem.* **2016**, *40*, 571–577. [CrossRef]

37. Jiang, J.; Ye, J.; Zhang, G.; Gong, X.; Nie, L.; Liu, J. Polymorph and Morphology Control of CaCO$_3$ via Temperature and PEG During the Decomposition of Ca(HCO$_3$)$_2$. *J. Am. Ceram. Soc.* **2012**, *95*, 3735–3738. [CrossRef]

38. Zheng, T.; Zhang, X.; Yi, H. Spherical vaterite microspheres of calcium carbonate synthesized with poly (acrylic acid) and sodium dodecyl benzene sulfonate. *J. Cryst. Growth.* **2019**, *528*, 125275. [CrossRef]

39. Qi, R.; Zhu, Y. Microwave-Assisted Synthesis of Calcium Carbonate (Vaterite) of Various Morphologies. *J. Phys. Chem. B* **2006**, *110*, 8302–8306. [CrossRef]

40. Nan, Z.; Chen, X.; Yang, Q.; Wang, X.; Shi, Z.; Hou, W. Structure transition from aragonite to vaterite and calcite by the assistance of SDBS. *J. Colloid. Interf. Sci.* **2008**, *325*, 331–336. [CrossRef]

41. Ramesh, T.N.; Inchara, S.A.; Pallavi, K. Para-amino benzoic acid-mediated synthesis of vaterite phase of calcium carbonate. *J. Chem. Sci.* **2015**, *127*, 843–848. [CrossRef]

42. Brennen, C.E. *Cavitation and Bubble Dynamics*; Cambridge University Press: Cambridge, UK, 1995. [CrossRef]

43. Azevedo, A.; Oliveira, H.; Rubio, J. Bulk nanobubbles in the mineral and environmental areas: Updating research and applications. *Adv. Colloid. Interf. Sci.* **2019**, *271*, 101992. [CrossRef] [PubMed]

44. Verwey, E.J.W.; Overbeek, J.T.G. Theory of the Stability of Lyophobic Colloids. *J. Colloid. Sci.* **1955**, *10*, 224–225. [CrossRef]

45. Derjaguin, B.; Landau, L. Theory of the Stability of Strongly Charged Lyophobic Sols and of the Adhesion of Strongly Charged Particles in Solutions of Electrolytes. *Prog. Surf. Sci.* **1993**, *43*, 30–53. [CrossRef]

46. Gogate, P.R.; Pandit, A.B. Engineering Design Methods for Cavitation Reactors II: Hydrodynamic Cavitation. *AIChE J.* **2000**, *46*, 1641–1649. [CrossRef]

47. Kontoyannis, C.G.; Vagenas, N.V. Calcium carbonate phase analysis using XRD and FT-Raman spectroscopy. *Analyst.* **2000**, *25*, 251–255. [CrossRef]

48. Proto, S.P.S.; Giordmaine, J.A.; Damen, T.C. Depolarization of Raman Scattering in Calcite. *Phys. Rev.* **1966**, *147*, 608–611. [CrossRef]

49. Nagabhushana, H.; Nagabhushana, B.M.; Lakshminarasappa, B.N.; Singhd, F.; Chakradhar, R.P.S. Swift heavy ion irradiation induced phase transformation in calcite single crystals. *Solid State Commun.* **2009**, *149*, 1905–1908. [CrossRef]

50. Elfil, H.; Roques, H. Role of hydrate phases of calcium carbonate on the scaling phenomenon. *Desalination* **2001**, *137*, 177–186. [CrossRef]

51. Zhang, F.; Yang, X.; Tian, F. Calcium carbonate growth in the presence of water soluble cellulose ethers. *Mater. Sci. Eng. C* **2009**, *29*, 2530–2538. [CrossRef]

52. Rumble, J.R. *CRC Handbook of Chemistry and Physics*, 98th ed.; CRC Press LLC.: Boca Raton, FL, USA, 2017.

53. Yoon, R.; Aksoy, B.S. Hydrophobic Forces in Thin Water Films Stabilized by Dodecylammonium Chloride. *J. Colloid. Interf. Sci.* **1999**, *211*, 1–10. [CrossRef] [PubMed]

54. Wang, L.; Yoon, R. Hydrophobic Forces in the Foam Films Stabilized by Sodium Dodecyl Sulfate Effect of Electrolyte. *Langmuir* **2004**, *20*, 11457–11464. [CrossRef] [PubMed]

55. Tao, D.; Wu, Z.; Sobhy, A. Investigation of nanobubble enhanced reverse anionic flotation of hematite and associated mechanisms. *Powder Technol.* **2021**, *379*, 12–25. [CrossRef]

56. Hampton, M.A.; Nguyen, A.V. Nanobubbles and the nanobubble bridging capillary force. *Adv. Colloid. Interf. Sci.* **2010**, *154*, 30–55. [CrossRef] [PubMed]

57. Tselishchev, Y.G.; Val'tsifer, V.A. Influence of the Type of Contact between Particles Joined by a Liquid Bridge on the Capillary Cohesive Forces. *Colloid. J.* **2003**, *65*, 385–389. [CrossRef]

58. Sirghi, L.; Szoszkiewicz, R.; Riedo, E. Volume of a Nanoscale Water Bridge. *Langmuir* **2006**, *22*, 1093–1098. [CrossRef]

59. Wang, X.; Yuan, S.; Liu, J.; Zhu, Y.; Han, Y. Nanobubble-enhanced flotation of ultrafine molybdenite and the associated mechanism. *J. Mol. Liq.* **2022**, *346*, 118312. [CrossRef]

60. Fan, M.; Tao, D.; Honaker, R.; Luo, Z. Nanobubble generation and its applications in froth flotation (part II): Fundamental study and theoretical analysis. *MST* **2010**, *20*, 159–177. [CrossRef]

61. Wang, D.; Liu, Q. Hydrodynamics of froth flotation and its effects on fine and ultrafine mineral particle flotation: A literature review. *Miner. Eng.* **2021**, *173*, 107220. [CrossRef]

62. Sobhy, A.; Tao, D. Nanobubble column flotation of fine coal particles and associated fundamentals. *Int. J. Miner. Process.* **2013**, *124*, 109–116. [CrossRef]

63. Li, B.; Zhang, H. A review of bulk nanobubbles and their roles in flotation of fine particles. *Powder Technol.* **2022**, *395*, 618–633. [CrossRef]

64. Priezjev, N.V.; Darhuber, A.A.; Troian, S.M. Slip behavior in liquid films on surfaces of patterned wettability: Comparison between continuum and molecular dynamics simulations. *Phys. Rev. E. Stat. Nonlin. Soft. Matter. Phys.* **2005**, *71*, 041608. [CrossRef] [PubMed]

 **materials**

Article

Simulation of Electrical and Thermal Properties of Granite under the Application of Electrical Pulses Using Equivalent Circuit Models

Kyosuke Fukushima [1], Mahmudul Kabir [1,*], Kensuke Kanda [1], Naoko Obara [1], Mayuko Fukuyama [2] and Akira Otsuki [3,4]

[1]   Graduate School of Engineering Science, Department of Mathematical Science and Electrical-Electronic-Computer Engineering, Tegata Campus, Akita University, 1-1 Tegata Gakuen Machi, Akita 010-8502, Japan; m8020422@s.akita-u.ac.jp (K.F.); m8021405@s.akita-u.ac.jp (K.K.); obara@gipc.akita-u.ac.jp (N.O.)

[2]   Graduate School of Engineering Science, Cooperative Major in Life Cycle Design Engineering, Tegata Campus, Akita University, 1-1 Tegata Gakuen Machi, Akita 010-8502, Japan; mayuko@gipc.akita-u.ac.jp

[3]   Ecole Nationale Supérieure de Géologie, GeoRessources, UMR 7359 CNRS, University of Lorraine, 2 Rue du Doyen, Marcel Roubault, BP 10162, 54505 Vandoeuvre-lès-Nancy, France; akira.otsuki@univ-lorraine.fr

[4]   Waste Science & Technology, Luleå University of Technology, SE 971 87 Luleå, Sweden

*   Correspondence: kabir@gipc.akita-u.ac.jp; Tel.: +81-18-889-2326

Abstract: Since energy efficiency in comminution of ores is as small as 1% using a mechanical crushing process, it is highly demanded to improve its efficiency. Using electrical impulses to selectively liberate valuable minerals from ores can be a solution of this problem. In this work, we developed a simulation method using equivalent circuits of granite to better understand the crushing process with high-voltage (HV) electrical pulses. From our simulation works, we calculated the electric field distributions in granite when an electrical pulse was applied. We also calculated other associated electrical phenomena such as produced heat and temperature changes from the simulation results. A decrease in the electric field was observed in the plagioclase with high electrical conductivity and void space. This suggests that the void volume in each mineral is important in calculating the electrical properties. Our equivalent circuit models considering both the electrical conductivity and dielectric constant of a granite can more accurately represent the electrical properties of granite under HV electric pulse application. These results will help us better understand the liberation of minerals from granite by electric pulse application.

Keywords: electric field; temperature distribution; conductivity; dielectric constant; liberation; hardrock; mineral distribution

Citation: Fukushima, K.; Kabir, M.; Kanda, K.; Obara, N.; Fukuyama, M.; Otsuki, A. Simulation of Electrical and Thermal Properties of Granite under the Application of Electrical Pulses Using Equivalent Circuit Models. *Materials* **2022**, *15*, 1039. https://doi.org/10.3390/ma15031039

Academic Editor: Georgios C. Psarras

Received: 31 December 2021
Accepted: 26 January 2022
Published: 28 January 2022

Publisher's Note: MDPI stays neutral with regard to jurisdictional claims in published maps and institutional affiliations.

Copyright: © 2022 by the authors. Licensee MDPI, Basel, Switzerland. This article is an open access article distributed under the terms and conditions of the Creative Commons Attribution (CC BY) license (https://creativecommons.org/licenses/by/4.0/).

1. Introduction

Minerals are familiar to us in our daily lives and are used in various fields (e.g., ceramics, pharmaceuticals, cosmetics) [1]. In order to extract minerals from rocks, a combination of mechanical comminution methods of crushing and grinding (e.g., ball mill) based on compression, impact and shearing mechanisms are generally used [2]. The energy efficiency of these methods is as low as 1%. Especially for hard rocks, such as granite, the mechanical grinding is inefficient due to tool wear and the increase in cost [3]. In addition, the energy used for comminution is very large, accounting for 2–4% of the world energy consumption [4,5]. Therefore, a new method for size reduction and selectively liberating minerals from rocks using high-voltage electric pulses has been developed [2]. Electric pulse liberation of minerals has been reported as a solution for low energy efficiency in comminution [6–8].

There are two types of electrical disintegration methods, and they are (1) Electro Hydraulic Disintegration (EHD) and (2) Electrical Disintegration (ED). Figure 1 shows a schematic of the EHD and ED methods. For (1) EHD, an upper/high voltage (HV) electrode and a ground electrode are placed above the object in a liquid and a shock wave is generated for the selective liberation of minerals (Figure 1a). (2) The ED method is also applied in a liquid but the upper/HV electrode is placed directly on or very close to the object to be disintegrated (Figure 1b). The ground electrode is placed below the object and when the electric pulse is applied, a large electric current is generated and flows through the object and thus leads to its disintegration. Water, due to its low electrical conductivity, is usually used as a liquid. For the EHD method, the voltage is applied only to the rock because the breakdown voltage of the rock is lower than the electric field strength of the water when the rise time of the electric pulse is less than 500 ns [9]. In the EHD method, lightning impulses are applied to two electrodes in close proximity, and the compressive force generated by the shock wave is used for crushing. In the ED method, the upper/HV electrode is placed close to the sample, and an instantaneous large current is applied to the inside of the sample to crush it [9]. The ED method is used in this study because of its potential for selective grinding and its ability to weaken the rock and reduce the energy consumption for grinding hard rocks [10,11].

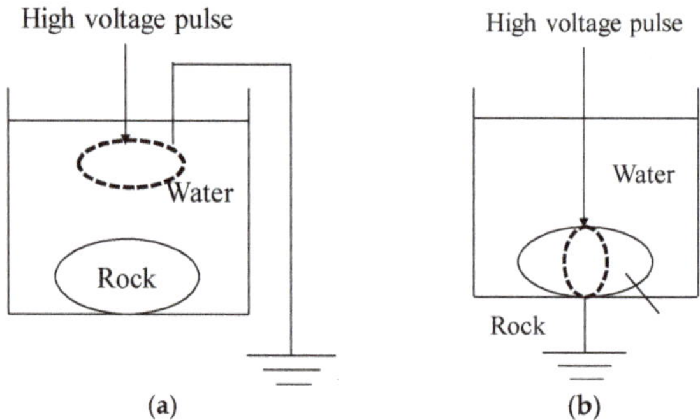

Figure 1. Model diagrams of different electric pulse crushing [4]. (**a**) EHD method; (**b**) ED method.

The ED method has been applied to the separation of precious minerals such as diamonds and emeralds [11]. In addition, the effects of electric pulse application on rocks have been studied based either on the electric conductivity or dielectric constant of rocks experimentally or simulation [11–13]. Walsh et al. (2020) performed simulation of hard rocks under HV pulse application considering the microstructures (i.e., distribution of minerals) of the rocks and they described about the change in minerals of the rocks under HV application. Li et al. (2018) showed how the *I-V* properties of granite changed under HV applications with different types of electrode spacing. However, the effects of electric pulses on rocks are still largely empirical, and our work is the very first study that considered both conductivity and dielectric constant as far as the authors are aware of.

The purpose of this study is to obtain better knowledge of granite comminution and selective liberation under high voltage pulses. Granite was selected as an example of hard rock. For this purpose, we developed an equivalent circuit model of granite based on both conductivity and dielectric constant, considering the minerals and voids in the granite, and applied a high electric field to the granite model to simulate electrical characteristics under HV applications. Specifically, our equivalent circuit model was created using a C# program, and the electric field distribution in granite was calculated using the circuit simulation software LTspice. Since rocks consist mostly of dielectric materials, our simulation works can be significantly helpful to understand the behavior of rocks (i.e., granite) under HV electric pulse, as the circuit models were proposed considering both electric and dielectric properties of granite.

2. Simulation Methods

2.1. Simulation Model

The simulation was performed for a well-known hard rock, i.e., granite composed of quartz, plagioclase, k-feldspar, and biotite [12]. In this study, a granite specimen with a cube of 1 mm on each side was considered, assuming that the entire rock is divided into an arbitrary number of smaller cubes to create an equivalent circuit model. In general, the dimension of a rock (i.e., granite in this work) is larger than that of our simulation works. We assumed the granite as a small cube of 1-mm size on each side in order to avoid the complexity of the work. As the electrical properties such as conductivity and relative permittivity do not vary with their sizes, it is quite reasonable to consider a smaller size for simulation works. By comparing larger and smaller sizes of the rocks, the results will show similar trends and this point was reconfirmed from our simulation results, similar to what was reported in the literature [12]. To obtain accurate results, we assumed 1000 parts of smaller cubes which make the cube of 1-mm edge length of granite, as shown in Figure 2a. The electrical properties of minerals can be represented by a parallel circuit of resistors and capacitors (Figure 2b). The resistance of minerals in the equivalent circuit model is calculated from the following equation:

$$R = \frac{1}{\sigma} \frac{L}{S}$$ (1)

where R is the resistance, σ is the electrical conductivity, L and S represent the length of the segmented mineral (0.1 mm) and the surface area of the segmented mineral (0.01 mm^2), respectively. Capacitance is expressed by the following equation:

$$C = \varepsilon \frac{S}{L}$$ (2)

here, C is capacitance and ε is dielectric constant. The equivalent circuit was created by assigning the minerals contained in the granite to the divided cubes, calculating the resistance and capacitor values for each mineral, and constructing the circuit. As an example, Figure 2 shows the equivalent circuit model for a $10 \times 10 \times 10$ partitioned cube. The simulation work was performed for a needle-flat electrode as shown in Figure 2a. From the knowledge of electric circuits, it can be understood that the electric field strength is mainly applied in the vertical direction. The electric field strength in the horizontal direction was neglected because the values were less than 3% compared to the ones in vertical direction. Thus, in this paper, we have shown the electric field strengths in the vertical direction only. The potential difference for each circuit element was calculated by dividing the values by the distance (i.e., 0.1 mm). Then, the electric field was calculated for each smaller cube and the data were used to draw the 3D mapping of the electric field distribution with MATLAB (R2016a, The MathWorks, Inc., Natick, MA, USA). Other simulation results were also used to draw 3D images by MATLAB.

2.2. Composition of Granite

The mineral composition of a typical granite is referred to the information available in [9], as shown in Table 1. The distribution of each mineral in the granite was determined by a random placement program based on the compositional ratio using the programming software C#.

Table 1. Mineral composition of a typical granite [12].

Mineral	Volume [%]
Quartz	30
Plagioclase	20
K-Feldspar	45
Biotite	5

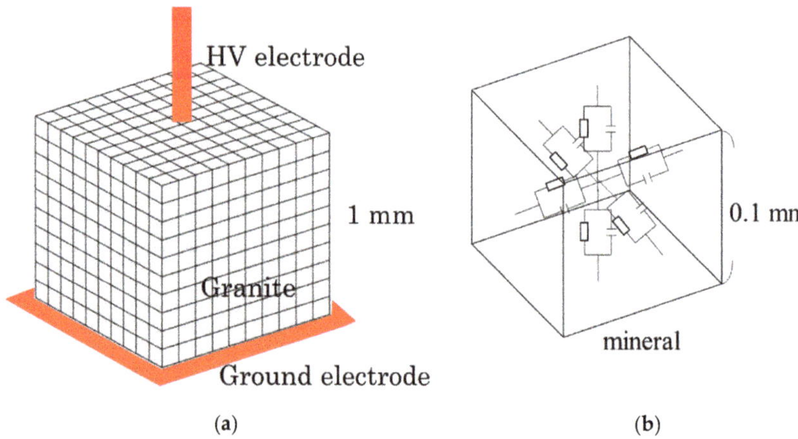

Figure 2. Concept of equivalent circuit model. (**a**) Granite; (**b**) equivalent circuit model for a small divided cube of granite.

2.3. Electrical and Thermal Properties of Minerals

The current J per unit area flowing through a rock is expressed by Ohm's law:

$$J = \sigma E + \frac{\partial D}{\partial t} = \sigma E + \varepsilon \frac{\partial E}{\partial t} \tag{3}$$

where E is the electric field, and ε is the dielectric constant [14]. From Ohm's law, the current distribution in a rock is affected by the electrical conductivity and dielectric constant. The relationship between temperature and electrical conductivity of minerals (σ_s) is empirically expressed by the following equation [12]:

$$\sigma_s = A\,exp\left(\frac{-B}{k_b T}\right) \tag{4}$$

where A and B are constants, T is the absolute temperature, and k_b is Boltzmann's constant (8.618×10^{-5} eV/K). The constants A and B of the granite rocks are shown in Table 2 [12], and the electrical conductivity of the minerals as a function of temperature is shown in Figure 3. The values of conductivity in Figure 3 were calculated for each single mineral using Equation (4). It is worth mentioning that the conductivity values of minerals (i.e., Figure 3) are surely different from the conductivity values of a mix solution (i.e., for NaCl solution of Equation (7)). At low temperature, the electrical conductivity of minerals is mainly affected by the water content in the fractures and pores. Sinmyo and Keppler reported the following empirical equation for NaCl solutions [15]:

$$\log_{10} \sigma_{Brine} = -1.7060 - \frac{93.78}{T} + 0.8075 \log_{10} c + 3.0781 \log_{10} \rho + \log_{10}(\lambda_0(\rho, T)) \tag{5}$$

where c is the mass percent concentration of NaCl, ρ is the mineral density, and λ_0 is the molar conductivity of NaCl in water at unlimited dilution. Here, λ_0 is approximated by the following equation:

$$\lambda_0 = 1573 - 1212\rho + \frac{537,062}{T} - \frac{208,122,721}{T^2} \tag{6}$$

Equation (5) can only be applied to temperatures above 100 °C, but the following equation can be used to approximate the electrical conductivity below 150 °C [16]:

$$\sigma_f = \sigma_{ref}\frac{T - 251.65}{T_{ref} - 251.65} \tag{7}$$

where σ_{ref} is the electrical conductivity at the reference temperature, and the unit of σ_{ref} is K. In this simulation, the σ_{ref}, T and c was set to 150 °C (423.15 K), 20 °C (293.15 K), and 3 wt%, respectively [12,15]. Note that the temperature variation occurs instantaneously and we used Equation (17) to calculate the temperature of the granite under HV impulse application.

Table 2. Values of the empirical parameters [12].

Mineral	Log(A) [log(s/m)]	B [eV]
Quartz	6.3	0.82
Plagioclase	0.041	0.85
K-Feldspar	0.11	0.85
Biotite	−13.8	0.00

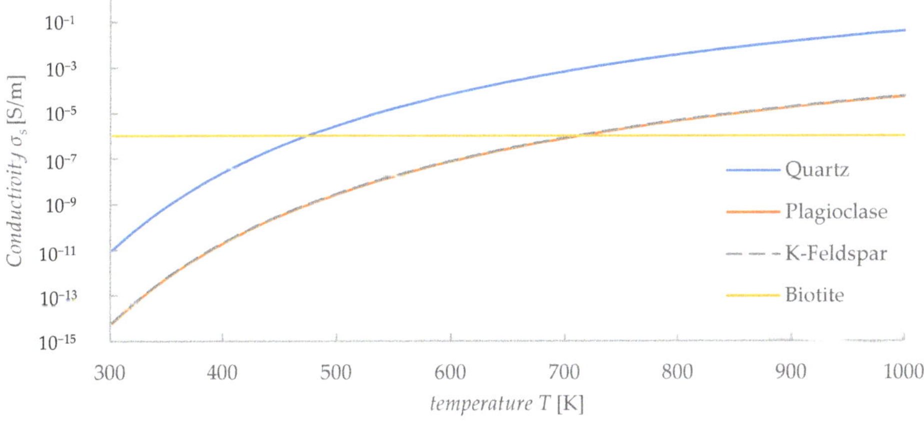

Figure 3. Calculated values of conductivity of minerals versus temperature.

The electrical conductivity for a composite mixture of solid and fluid can be expressed using a modified Archie's law proposed by Glover [17]:

$$\sigma_{mix} = \sigma_s\left(1 - \varnothing_f\right)^p + \sigma_f\varnothing_f{}^m \tag{8}$$

where σ_s is the electrical conductivity of a solid, σ_f is the electrical conductivity of a fluid, $\varnothing_f$ is the volume fraction of a fluid, m is Archie's exponent, and p is determined by Equation (9).

$$p = \frac{\log\left(1 - \varnothing_f{}^m\right)}{\log\left(1 - \varnothing_f\right)} \tag{9}$$

The value of Archie's index (m = 1.5) was used in this calculation and obtained from [12].

As shown in Equation (1), the value of dielectric constant is important in considering the current distribution within the minerals composing a rock. The volume of the fluid (considered equivalent to the void volume) and the dielectric constant of each mineral are shown in Table 3 [12,13].

Table 3. Values of void volume * and permittivity of minerals in granite [12,13].

Mineral	Void [%]	Permittivity
Quartz	0.9	6.53
Plagioclase	1.8	6.91
K-Feldspar	0.9	6.2
Biotite	0.9	9.28

* Volume filled with water during ED method application.

2.4. Applied Voltage

In the case of ED crushing, the electrodes are placed directly on the rock and the voltage is applied, and thus, the mineral crushing is theoretically possible even without using lightning impulses. However, in practice, the application of high voltage for a long period needs a large amount of electricity and lowers the energy efficiency. Due to this point, it is reasonable to use lightning impulses for mineral crushing. A lightning impulse (0.3 μs/50 μs) was used as the input voltage applied to the granite in this work. The peak value of impulse voltage was at the front time (i.e., 0.3 μs = 300 ns), and 50% of the peak voltage was at the tail time (i.e., 50 μs). In this work, the lightning impulse was simulated and applied to the granite by the circuit simulator LTspice (Figure 4). Fujita et al. (1999) explained that the voltage was applied only to the rock when the rise time of the applied voltage was less than 500 ns in the ED method [9]. Therefore, in this study, the applied voltages were: (a) 100% rise, the maximum voltage 20 kV at a rise time of 300 ns, and (b) 50% rise, 50% of the maximum voltage 10 kV at 38 ns. The electric current I was also calculated and plotted within the same graph of applied voltage (i.e., Figure 4). The maximum current (i.e., 2.55 mA) was observed at 2.34 ps. At the maximum value of the lightning impulse voltage (i.e., 20 kV), the value of the current was 1.53 mA at 0.3 μs. After 0.3 μs, the value of the current gradually decreased to 0.75 mA at 50 μs.

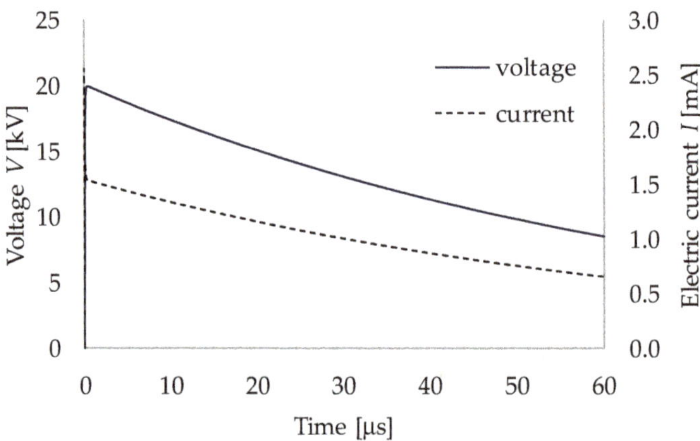

Figure 4. Applied voltage used in this simulation work and electric current.

2.5. Temperature in the Rock

Assuming a rock is regarded as a dielectric material, we will discuss the electric phenomena when an AC input voltage v with angular frequency ω is applied on the rock. Figure 5 shows the equivalent circuit for a typical dielectric material with AC input voltage. The AC analysis for this circuit can be explained with the use of vector analysis. The input voltage was chosen as the reference vector. The effective value of voltage v is V and the vector expression we used is the symbol of $\dot{V}$. For the circuit shown in Figure 5a, the total electric current $\dot{I}_p$ can be expressed by a vector sum of $\dot{I}_C$ and $\dot{I}_R$. Here, $\dot{I}_C$ represents the displacement current and $\dot{I}_R$ represents the loss current of the granite. $\dot{I}_C$ and $\dot{I}_R$ can be explained by Equations (10) and (11), respectively.

$$\dot{I}_C = j\omega C\dot{V} \tag{10}$$

$$\dot{I}_R = \frac{1}{R}\dot{V} \tag{11}$$

Here, ω is angular frequency of the AC voltage v. The sum of the electric current, i.e., $\dot{I}_P$ can be found by adding Equations (10) and (11).

$$\dot{I}_P = \dot{I}_R + \dot{I}_C = \left(\frac{1}{R} + j\omega C\right)\dot{V} \tag{12}$$

The amount of heat per unit time generated by the applied voltage and current flowing through the granite (i.e., electric power) can be calculated by the following equation:

$$\frac{\partial q}{\partial t} = VI_P\cos\theta = VI_R = \frac{1}{R}V^2 \tag{13}$$

where q is the heat generated, V is the potential difference and θ is the phase angle between the voltage and electric current $\dot{I}_P$. Assuming the phase angle between $\dot{I}_C$ and $\dot{I}_P$ is δ, the relationship in between θ and δ can be expressed by the following equation,

$$\theta + \delta = \frac{\pi}{2} \tag{14}$$

From Figure 5b, the dielectric loss $\tan\delta$ can be expressed by the following equation.

$$\tan\delta = \frac{\dot{I}_R}{\dot{I}_C} = \frac{1}{R\omega C} \tag{15}$$

The values of resistance R and capacitance C were calculated for each mineral as shown in Table 4. The $\tan\delta$ was also calculated at 20 kHz as shown in Table 4. From Equation (15), it is understood that the dielectric loss (i.e., $\tan\delta$) changes with frequency. The value of $\tan\delta$ was 175 for quartz (see Table 4), but it can be changed to 1.17 at the frequency of 3 MHz. The temperature change caused by this heat can be expressed using a thermal resistance defined by the following equation:

$$R_j = \frac{1}{k}\frac{L}{S} \tag{16}$$

where R_j is the thermal resistance and k is the thermal conductivity. The relationship between temperature change and heat is expressed by the following equation:

$$\frac{\partial q}{\partial t} = \frac{T_1 - T_2}{R_j} \tag{17}$$

where T_1 and T_2 are the temperatures before and after the application of high voltage (i.e., heat generation) to the granite, respectively. The thermal conductivity of each mineral is summarized in Table 5 [18]. Equation (17) was used to calculate the temperature of granite under the HV application to be discussed later.

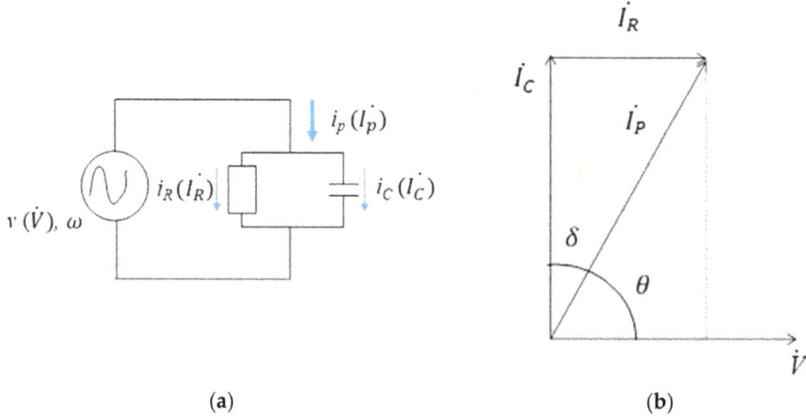

Figure 5. Equivalent circuit model for typical dielectric material. (**a**) Equivalent circuit; (**b**) vector analysis of the circuit.

Table 4. Calculated values of resistance R, capacitance C and $\tan\delta$ of minerals in granite.

Mineral	Resistance R [Ω]	Capacitance C [F]	$\tan\delta$
Quartz	1.57×10^7	2.90×10^{-15}	175
Plagioclase	5.54×10^6	3.08×10^{-15}	467
K-Feldspar	1.57×10^7	6.15×10^{-15}	82.5
Biotite	1.57×10^7	8.26×10^{-15}	61.4

Table 5. Heat characteristics of minerals in granite [18].

Mineral	Heat Resistance [K/W]	Heat Conductivity [W/m·K]
Quartz	1300	7.69
Plagioclase	4673	2.14
K-Feldspar	4329	2.31
Biotite	4950	2.02

3. Results and Discussion

3.1. Distribution of Minerals in Granite

Using a program developed in C#, the distribution of minerals was determined, based on the mineral composition of the granite shown in Table 1. The layout information of the granite rock is shown in Figure 6a,b. Since the placement information is three-dimensional, we displayed in a cross-sectional view where the electrode is assumed to be placed, as shown in Figure 6b that is also a reference to Figure 6c–h in terms of mineral and electrode placement. The mineral distribution was chosen randomly meeting the conditions of Tables 1 and 3. Some works [19,20] reported mineral distributions arranged with Voronoi networks in order to describe any random distribution of particles, minerals, etc. Though our simulation works used a 5 times larger void, comparing to their works, the proportion of void and minerals was maintained as described in Table 3.

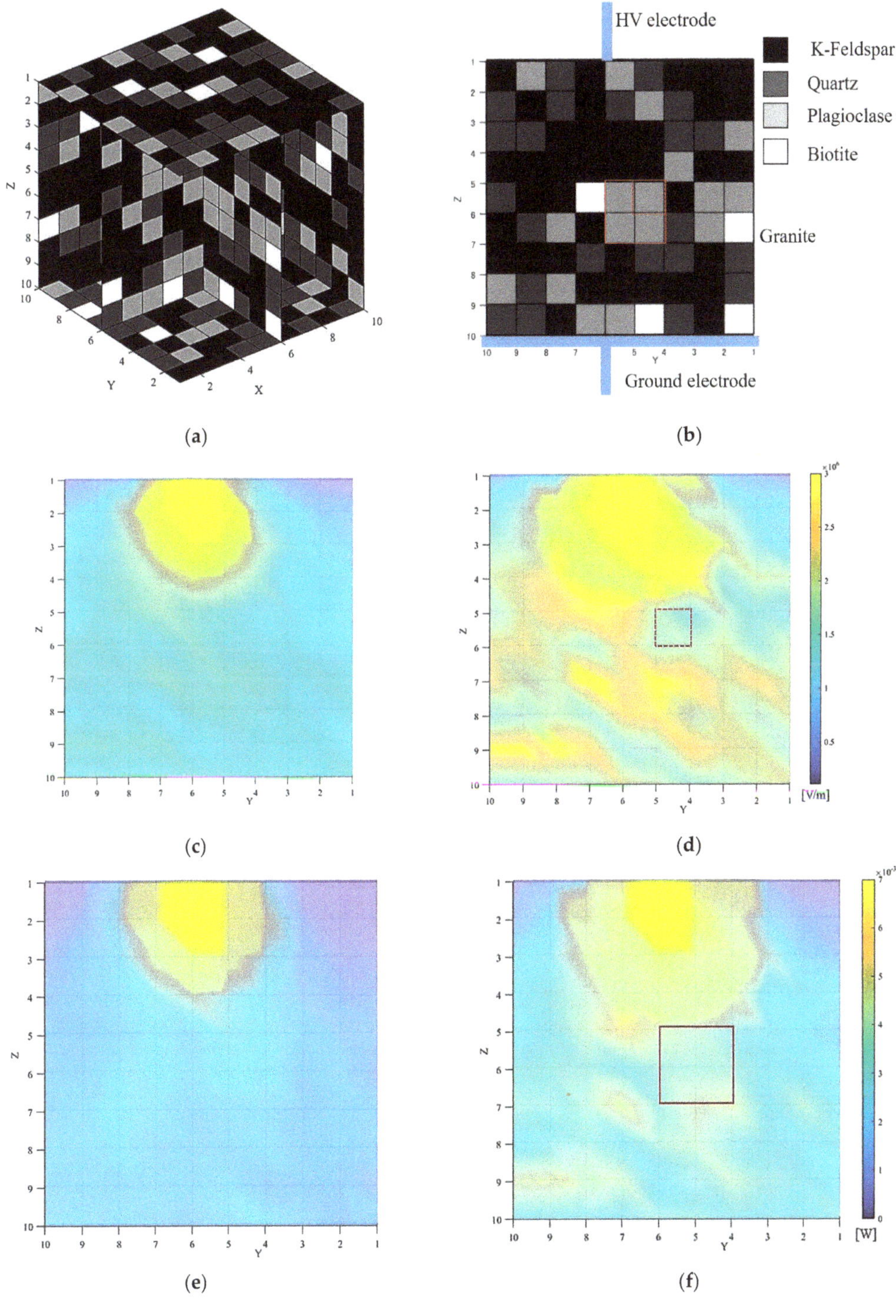

(**a**)

(**b**)

(**c**)

(**d**)

(**e**)

(**f**)

Figure 6. *Cont.*

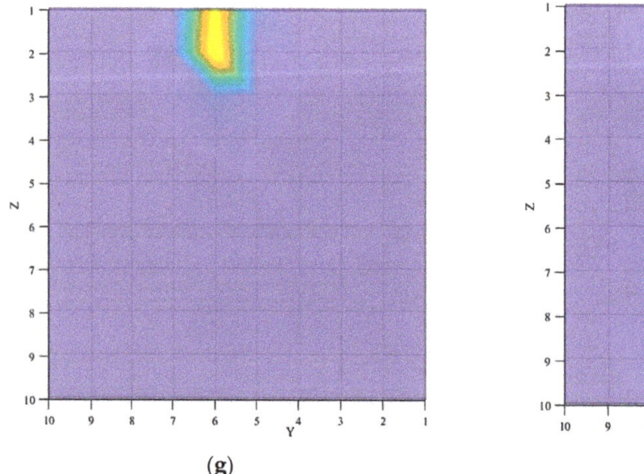
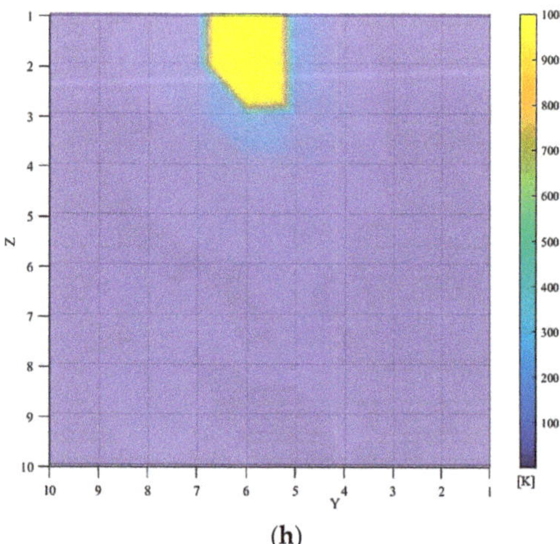

(g)　　　　　　　　　　　　　　　(h)

Figure 6. Simulation result when time changes. The red boxes in (**b,d,f**) are added for the visual aid. (**a**) Mineral distribution; (**b**) cross-section surface viewed from y-z direction; (**c**) electric field (50% rise of lightning impulse voltage); (**d**) electric field (100% rise of lightning impulse voltage); (**e**) heat (50% rise of lightning impulse voltage); (**f**) heat (100% rise of lightning impulse voltage); (**g**) variation of temperature (50% rise of lightning impulse voltage); (**h**) variation of temperature (100% rise of lightning impulse voltage).

3.2. Simulation Results

The rock was divided into 1000 small blocks (10 × 10 × 10), and minerals were placed in the blocks based on the mineral composition of the granite rock (Table 1). An equivalent circuit model was created based on the mineral distribution information (Figure 6b), and was used to investigate the electric field distribution in the granite when an electric pulse was applied to the granite. The applied voltage was set to reach the maximum voltage 20 kV at around 300 ns, as shown in Figure 4. The heat distribution in the granite was calculated from the electric field and the value of resistors. The temperature change was calculated from the heat distribution and the values of thermal resistance in the granite. The results of each calculated parameter for 20 kV (100% rise) and 10 kV (50% rise) are summarized in Figure 6. Table 6 shows the maximum and minimum values for each result. All of the maximum values in Table 6 were found in the small block in contact with the upper/HV electrode, and the values decreased as they moved away from the upper/HV electrode, with the minimum value found in the area in contact with the ground electrode.

Table 6. Maximum and minimum values for each result.

Voltage	Maximum Value of Electric Field [V/m]	Minimum Value of Electric Field [V/m]	Maximum Value of Heat [W]	Minimum Value of Heat [W]	Maximum Value of Temperature Change [K]	Minimum Value of Temperature Change [K]
50% rise	17.7×10^6	106×10^3	0.262	0.00	1133	0.0283
100% rise	29.7×10^6	104×10^3	0.7351	0.00	3182	0.0664

Figure 6c shows the electric field distribution at the moment when 50% rise voltage was applied, and Figure 6d shows the electric field distribution at the moment when 100% rise voltage was applied. As mentioned above, the electric field distribution was high near the upper/HV electrode. In addition, the electric field distribution was found lower inside the plagioclase compared to other minerals. In Figure 6d, for example, plagioclase placed in the area defined by the y-z coordinates (4,5), (4,6) (5,6), (5,5) showed a lower electric field of plagioclase compared to their surroundings. Plagioclase has a higher void volume than other minerals (1.8 > 0.9%) and contains more water in the ED method (see Table 3) and that can be the reason behind the lower electric field distribution in plagioclase. In this study, the initial temperature of the minerals was set to 20 °C, so the electrical conductivity of the mineral-water composite was highly dependent on the water content. In other words, as the amount of water increases, the electrical conductivity increases (Equation (8)). Since electrical conductivity and resistivity have an inverse relationship, the resistance decreases, and thus, the electric field decreases according to Equation (3). This might be the reason for the low electric field inside plagioclase. When the resistance of the minerals adjacent to the plagioclase is lower than that of plagioclase, the voltage of the plagioclase with lower resistance is inevitably lower than that of the surrounding area because the current flowing in the circuit branches in the vertical direction is the same. Since the electric field is calculated by dividing the potential difference between two adjacent contacts by the distance between the two contacts, the electric field of the plagioclase naturally becomes lower than the surrounding area, while a higher electric field appears in the surrounding area. In comparing 50% between 100% rise of lightning impulse voltages (Figure 6c 50% rise vs. Figure 6d 100% rise), the latter case shows a clearer difference in the electric field distribution possibly due to the change of the electric field distribution in the granite depending on the magnitude of the applied voltage. It can be due to the difference in the dielectric breakdown electric field between water and rock (minerals) and its threshold duration of the HV wavefront is about 500 ns [9]. In other words, below 500 ns, the current flows inside the minerals, otherwise it flows through the water and not through the minerals.

Figure 6e,f shows the heat distribution at the moment of applying 50% rise voltage and 100% rise voltage, respectively. We can see the change in the heat generated in the granite from Figure 6e,f calculated from Equation (13). The heat generated in the upper part of the granite was higher than that of the lower part and was similar to the electric field distribution. There was almost no heat generated in the bottom parts in granite from our simulations, especially for the 50% rise of lightning impulse voltage (Figure 6e). As shown in Equation (13), the amount of heat generated is determined by the electric field distribution and the value of resistance, and since it is known from Figure 6c,d that the resistance of plagioclase is small due to its high electric conductivity, the electric field is low. Figure 6g,h shows the temperature change calculated at 50% rise and 100% rise of lightning impulse voltage. The maximum temperature change was about 3182 K near the upper/HV electrode, but the temperature change near to the ground electrode was almost zero. In particular, the plagioclase present in the area defined by the y-z coordinates (4,5), (4,7) (6,7), (6,5) showed less heat generation than their surrounding areas which showed a relatively high heat generation compared to plagioclase. The results showed that the temperature change differs with minerals and distance from the upper/HV electrode. By comparing Figure 6e (50% rise) with Figure 6f (100% rise), the temperature distribution is more heterogeneous in the latter case and affected by the different mineral placement that can deviate the electric field distribution (Figure 6d). Again, this temperature distribution difference can be the difference in the dielectric breakdown electric field as in the case of the electric field distribution.

As seen in Figure 6g,h, the temperature change is larger in the vicinity of the upper/HV electrode, indicating that the temperature change was occurring more likely by the heat generated due to the electric field and the resistance value of each mineral rather than by the heat resistance. The results obtained in this study reproduced similar trends as reported in other research works where the calculation was performed by considering only the electrical conductivity of minerals [12,13].

Walsh et al. (2013) used Voronoi tessellation to simulate granite in order to understand the thermal spallation. Their works using the Euler–Godunov model showed that the heat distribution differs in the boundary lines of the Voronoi tessellation (i.e., minerals). Our simulation works also showed that in spite of electric field distribution differences in the granite from their work, the heat variation was very small (see Figure 6f) [19].

Walsh et al. (2020) described simulation results with a 10-cm wide granite model with thickness of 2 cm. We used the same mineral composition in the granite as they did. They performed electric pulse crushing simulation using the ED method with needle electrodes placed at the center of the granite model for both the upper/HV and ground electrodes in 2D [12]. In addition, a high electric field was observed around the electrodes. In comparison with our results, they also observed a decrease in the electric field at the plagioclase and high electric field distribution near the electrodes [12]. In our work, we used both electric and dielectric parameters to understand the behavior of granite under HV electric pulse application. From our work, as we described before, the electric properties affected the electric distribution and heat generated in the minerals. At the same time, by considering dielectric constant in our work, the reason of the generated heat difference in different minerals was clear. On the other hand, Walsh et al. (2020) used 10 times larger input voltage compared to our work in order to simulate the thermal differences. If we consider the economic side of the mineral liberation by electric pulses, our method can be used to understand electrical and thermal properties of minerals under a lower input voltage application. Within this framework, further simulation works are to be carried out to confirm the heat difference under larger HV applications.

An application example of the ED method was described in the literature where it was applied on several minerals (i.e., quartz, pyrite, calcite, albite) which were embedded in cement and their selective liberation behavior was observed [19]. The size of the cemented minerals was 10 mm × 10 mm × 8 mm and the HV electrode was a needle electrode and flat electrode was placed as a grounding electrode. Channels were formed near the contact point of the electrodes, and crushing occurred when the electric pulse was applied on the cemented minerals [21]. The formation of channels and their connections which led to crushing occurred near the needle of the electrode and was due to the high electric field generated by the electric pulse. For the sample of quartz and cement, crushing occurred along the boundary between quartz and cement. The electric field distribution changed at the boundary of cement and quartz due to the difference in their electric conductivities, and this phenomenon is similar to the relationship between plagioclase and other minerals found in our study (Figure 6d,f). The electric field data obtained in this study are consistent with the general electric-pulse comminution theory, which states a difference in electric field occurs at boundaries with large differences in electric conductivity and dielectric constant, causing rock crushing due to rapid expansion of the discharge channels and associated heat generation [2]. The above comparison with the literature indicated that our simulation results showed the same/similar trend to the previous simulation and experiments of applying ED method on rocks, and thus, our methods and results obtained in this study are considered valid.

4. Conclusions

In this study, we investigated the electrical and thermal characteristics of granite under the application of electric pulses. An equivalent circuit model of a granite in 3D (1 mm × 1 mm × 1 mm) was developed. The electric field in the circuit was then calculated considering an electric pulse application. The heat and temperature changes generated in the minerals with the granite were calculated from that result. Until now, existing studies used either electrical conductivity or dielectric constant, but there has been no model that considers both. In this study, however, we used both parameters in the equivalent circuit model, calculated the values of the electric field distribution, and then, the heat distribution, and created the temperature change distribution from the heat distribution. This has not been performed before, and can lead to more detailed data under relatively low applied voltage by considering both parameters; so, the findings of this study can be useful for energy efficient rock crushing. The results of this study are summarized as follows:

(1) The electric field of the granite confirmed that the electric field dropped near the plagioclase;

(2) The electrical conductivity of the plagioclase was higher than that of the other minerals due to its higher void volume, which has a significant effect on the electric field distribution;

(3) The temperature change in the granite observed a very high temperature change near the upper/HV electrode, but it was very small as away from that electrode.

Author Contributions: Conceptualization, M.K. and A.O.; methodology, K.F., M.K., K.K., N.O., M.F., A.O.; software, K.F., M.K., K.K.; validation, K.F., M.K., K.K. and A.O.; formal analysis, K.F., M.K., K.K.; investigation, K.F., M.K., K.K.; resources, M.K. and A.O.; data curation, K.F., M.K., K.K. and A.O.; writing—original draft preparation, K.F., M.K.; writing—review and editing, K.F., M.K., A.O.; visualization, K.F., M.K., K.K.; supervision, M.K. and A.O.; project administration, M.K. and A.O.; funding acquisition, M.K. and A.O. All authors have read and agreed to the published version of the manuscript.

Funding: This research received no external funding.

Institutional Review Board Statement: Not applicable.

Informed Consent Statement: Not applicable.

Data Availability Statement: Not applicable.

Conflicts of Interest: The authors declare no conflict of interest.

References

1. Otsuki, A.; De Campo, L.; Garvey, C.J.; Rehm, C. H_2O/D_2O Contrast Variation for Ultra-Small-Angle Neutron Scattering to Minimize Multiple Scattering Effects of Colloidal Particle Suspensions. *Colloids Interfaces* **2018**, *2*, 37. [CrossRef]

2. Andres, U.; Timoshkin, I.; Jirestig, J.; Stallknecht, H. Liberation of valuable inclusions in ores and slags by electrical pulses. *Powder Technol.* **2001**, *114*, 40–50. [CrossRef]

3. Tester, J.W.; Anderson, B.J.; Batchelor, A.S. Impact of enhanced geothermal systems on US energy supply in the twenty-first century. *Phil. Trans. Math. Phys. Eng. Sci.* **2007**, *365*, 1057–1094. [CrossRef] [PubMed]

4. Liberation of Minerals with Electric Pulse (Digital Description of Prof. Shuji Owada's Work, Waseda University, Japan). Available online: https://www.waseda.jp/top/news/6137 (accessed on 20 December 2021). (In Japanese)

5. Taylor, L.; Skuse, D.; Blackburn, S.; Greenwood, R. Stirred media mills in the mining industry: Material grindability, energy-size relationships, and operating conditions. *Powder Technol.* **2020**, *369*, 1–16. [CrossRef]

6. Andres, U. Development and prospects of mineral liberation by electrical pulses. *Int. J. Miner. Process.* **2010**, *97*, 31–38. [CrossRef]

7. Usov, A.F.; Tsukerman, V.A. Prospective of electric impulse processes for the study of the structure and processing of mineral raw materials. *Dev. Miner. Process.* **2000**, *13*, C2-8–C2-15.

8. Gao, P.; Yuan, S.; Han, Y.; Li, Y.; Chen, H. Experimental study on the effect of pretreatment with high-voltage electrical pulses on mineral liberation and separation of magnetite ore. *Minerals* **2017**, *7*, 153. [CrossRef]

9. Fujita, T.; Yoshimi, I.; Tanaka, Y.; Jeyadevan, B.; Miyazaki, T. Research of liberation by Using High Voltage Discharge Impulse and Electromagnetic Waves. *Min. Mater. Process. Inst. Jpn.* **1999**, *115*, 749–754, (In Japanese with English Abstract).

10. Andres, U.; Bialecki, R. Liberation of mineral constituents by high-voltage pulses. *Powder Technol.* **1986**, *48*, 269–277. [CrossRef]

11. Fujita, T.; Shibayama, A. Liberation for the pre-treatment of recycle by means of electrical disintegration and disintegration by underwater explosion. *Resour. Process.* **2002**, *49*, 187–196. [CrossRef]

12. Walsh, S.D.C.; Vogler, D. Simulating electropulse fracture of granitic rock. *Int. J. Rock Mech. Min. Sci.* **2020**, *128*, 104238. [CrossRef]

13. Li, C.; Duan, L.; Tan, S.; Chikhotkin, V. Influences on High -Voltage Electro Pulse Boring in Granite. *Energies* **2018**, *11*, 2461. [CrossRef]

14. Nishizawa, O.; Kanagawa, K. *Introduction to the Physics of Rocks*; Maruzen Publishing: Tokyo, Japan, 2012.

15. Sinmyo, R.; Keppler, H. Electrical conductivity of NaCl-bearing aqueous fluids to 600 °C and 1 gpa. *Contrib. Mineral. Petrol.* **2017**, *172*, 4. [CrossRef]

16. Arps, J.J. The effect of temperature on the density and electrical resistivity of sodium chloride solutions. *J. Petrol. Technol.* **1953**, *5*, 17–20. [CrossRef]

17. Glover, P.W.; Hole, M.J.; Pous, J. A modified Archie's law for two conducting phases. *Earth Planet. Sci. Lett.* **2000**, *180*, 369–383. [CrossRef]

18. Vogler, D.; Walsh, S.D.C.; Rudolf von Rohr, P.; Saar, M.O. Simulation of rock failure modes in thermal spallation drilling. *Acta Geotech.* **2020**, *15*, 2327–2340. [CrossRef]

19. Walsh, S.D.C.; Lomov, I.N. Micromechanical modeling of thermal spallation in granitic rock. *Int. J. Heat Mass Tran.* **2013**, *65*, 366–373. [CrossRef]

20. Singh, J.P.; Walsh, S.D.C.; Koch, D.L. Brownian dynamics of a suspension of particles with constrained voronoi cell volumes. *Langmuir* **2015**, *31*, 6829–6841. [CrossRef]
21. Cho, S.H.; Yokota, M.; Ito, M.; Kawasaki, S.; Jeong, S.B.; Kim, B.K.; Kaneko, K. Electrical disintegration and micro-focus X-ray CT observations of cement paste samples with dispersed mineral particles. *Miner. Eng.* **2014**, *57*, 79–85. [CrossRef]

Article

Equivalent Circuit Models: An Effective Tool to Simulate Electric/Dielectric Properties of Ores—An Example Using Granite

Kyosuke Fukushima [1], Mahmudul Kabir [1,*], Kensuke Kanda [1], Naoko Obara [1], Mayuko Fukuyama [2] and Akira Otsuki [3,4,5]

[1] Graduate School of Engineering Science, Akita University, 1-1 Tegata Gakuen Machi, Akita 010-8502, Japan; m8020422@s.akita-u.ac.jp (K.F.); m8021405@s.akita-u.ac.jp (K.K.); obara@gipc.akita-u.ac.jp (N.O.)
[2] Graduate School of Engineering Science, Cooperative Major in Life Cycle Design Engineering, Tegata Campus, Akita University, 1-1 Tegata Gakuen Machi, Akita 010-8502, Japan; mayuko@gipc.akita-u.ac.jp
[3] Ecole Nationale Supérieure de Géologie, GeoRessources, UMR 7359 CNRS, University of Lorraine, 2 Rue du Doyen, Marcel Roubault, BP 10162, 54505 Vandoeuvre-lès-Nancy, France; akira.otsuki@uai.cl
[4] Facultad de Ingeniería y Ciencias, Universidad Adolfo Ibáñez, Diagonal Las Torres 2640, Peñalolén, Santiago 7941169, Chile
[5] Waste Science & Technology, Luleå University of Technology, SE 971 87 Luleå, Sweden
* Correspondence: kabir@gipc.akita-u.ac.jp; Tel.: +81-18-889-2326

Citation: Fukushima, K.; Kabir, M.; Kanda, K.; Obara, N.; Fukuyama, M.; Otsuki, A. Equivalent Circuit Models: An Effective Tool to Simulate Electric/Dielectric Properties of Ores—An Example Using Granite. *Materials* **2022**, *15*, 4549. https://doi.org/10.3390/ma15134549

Academic Editor: Saeed Chehreh Chelgani

Received: 8 April 2022
Accepted: 24 June 2022
Published: 28 June 2022

Publisher's Note: MDPI stays neutral with regard to jurisdictional claims in published maps and institutional affiliations.

Copyright: © 2022 by the authors. Licensee MDPI, Basel, Switzerland. This article is an open access article distributed under the terms and conditions of the Creative Commons Attribution (CC BY) license (https://creativecommons.org/licenses/by/4.0/).

Abstract: The equivalent circuit model is widely used in high-voltage (HV) engineering to simulate the behavior of HV applications for insulation/dielectric materials. In this study, equivalent circuit models were prepared in order to represent the electric and dielectric properties of minerals and voids in a granite rock sample. The HV electric-pulse application shows a good possibility of achieving a high energy efficiency with the size reduction and selective liberation of minerals from rocks. The electric and dielectric properties were first measured, and the mineral compositions were also determined by using a micro-X-ray fluorescence spectrometer. Ten patterns of equivalent circuit models were then prepared after considering the mineral distribution in granite. Hard rocks, as well as minerals, are dielectric materials that can be represented as resistors and capacitors in parallel connections. The values of the electric circuit parameters were determined from the known electric and dielectric parameters of the minerals in granite. The average calculated data of the electric properties of granite agreed with the measured data. The conductivity values were 53.5 pS/m (measurement) and 36.2 pS/m (simulation) in this work. Although there were some differences between the measured and calculated data of dielectric loss ($tan\delta$), their trend as a function of frequency agreed. Even though our study specifically dealt with granite, the developed equivalent circuit model can be applied to any other rock.

Keywords: conductivity; dielectric constant; hard rock; mineral distribution; voltage-dependent resistance (VDR)

1. Introduction

The modeling of rocks is one of the effective tools available to understand the physical phenomena in rocks under different applications, including the electric-pulse liberation of minerals. The electric-pulse liberation of minerals is an emerging comminution method that can be a good solution for low-energy efficiency in size reduction and selective liberation [1,2]. As rocks are dielectric materials [3–6], it is necessary to understand their electric and dielectric properties to evaluate and better understand the physics and behaviors of minerals upon the application of a HV electric pulse on them. Since the 1990s, computer simulations have become an important tool to model the research due to the unprecedented development of the data processing power of personal computers. There are many works related to electric-pulse liberation being carried out using computer simulations in order to understand the HV electric-pulse liberation of rocks [7].

Andres et al. (2001) compared the electrical pulse application with the conventional mechanical comminution of oxide ores containing hematite or platinum-group metals [1]. They also showed some simulation works of the electric-field distribution in ores using the electromagnetic theory of dielectric materials.

Some other works were performed to understand the electric-pulse comminution by using the COMSOL Multiphysics software package [8,9]. Seyed et al. (2015) worked on phosphate ore that was under electric-pulse application, which showed that the electrical field was dependent on the electrical properties of minerals, particle size and the location of conductive minerals [8]. Li et al. (2018) used the COMSOL Multiphysics software to understand the behavior of electro-pulse boring in granite under a HV application, considering the composition of granite, electrode spacing and electrode shape. They found that HV boring is affected by the composition of granite and its electric properties. However, they used equivalent circuit models of an electric pulse of a HV source only [9].

Zuo et al. (2015) discussed high-voltage pulse (HVP) breakage models of ores with three breakage indices (i.e., body breakage probability, body reduction evaluation index, body breakage product pre-weakening degree) [10]. They considered the breakage results of three ore samples (i.e., gold–copper ore, iron oxide copper–gold ore (IOCG), and hematite ore) and fitted the measured data statistically. Their work linked these parameters with the mass-specific impact energy, and even though the ores were different, the HVP model showed similar results with measured data [10].

Walsh and his groups conducted some simulation works related to the HV pulse liberation of ores [11,12], using Voronoi tessellation in order to simulate the mineral distribution in the granite rocks. Their works considered the minerals of granite to understand their behaviors under HV electric applications. However, these simulation works did not consider both the electric and dielectric properties of each mineral consisting of rock.

In HV engineering, equivalent circuit models that represent insulation and dielectric materials are used to simulate and understand the behavior of HV in testing objects [9,13–17]. Zuo et al. (2020) modeled insulation cable with equivalent circuit models by considering placing the cable into small fractions of insulation (dielectric) material and discussed the HV direct current influence on insulation cables [14]. Kabir et al. (2011) [18] made equivalent circuit models to understand the fine ceramics (ZnO varistors) by considering hundreds of ZnO grains of an average grain size of 1–2 μm to understand the nonlinear I–V (current–voltage) properties of ZnO varistors. Ono et al. (2009) [17] calculated the electric-field distribution in composite materials using equivalent circuit models under HV applications. Their works discussed different types of electric pulses and electrode gaps to understand HV behavior. With the above literature, the importance of considering each mineral to understand mineral liberation under a HV application is clearly identified. Additionally, it is worth noting that the equivalent circuit model analysis has a strong potential to understand the electric pulse comminution [18].

The equivalent circuit model is used in HV applications, as discussed before. The model can simulate the behavior of the subject under a HV application effectively, but only if the equivalent circuit can imitate the subject's electrical/dielectric properties accurately. Thus, it is necessary to confirm the reproduction ability of the electrical/dielectric properties of a subject. However, the dielectric properties of composite materials are difficult to simulate by equivalent circuit models, as they are frequency-dependent and the dielectric values are fixed in an equivalent circuit model. We will discuss this regarding the simulation results.

We discussed the HV application on granite in the other paper [18], where an equivalent circuit was used to simulate the behavior of granite under a HV impulse application. As the results proved the effectiveness of our equivalent circuit model in electric-pulse liberation, in the current work, we aimed to further advance this model by considering voids' (pores in minerals as well as rocks) behavior under a HV impulse on the rocks by using the measurement data of electrical properties of the dielectric breakdown of air. Our previous paper also dealt with voids (pores in minerals), but we assumed those voids were filled with liquid, as the electrical disintegration (ED) method applies HV impulses in liquid. Thus, the values of electric resistance that include voids, did not represent the dielectric-breakdown properties of voids (pores) in minerals; rather, it showed the electric resistance of minerals including voids, as they were assumed to be filled with liquid. Generally, the electric and dielectric properties of any dielectric and insulation materials are measured in the HV branch in air [13]. In this paper, we considered all the voids (pores in minerals), and their equivalent circuits were placed along with the circuits of minerals in the equivalent circuit model of granite. Again, we considered both the electric and dielectric properties of each type of mineral of granite (i.e., quartz, plagioclase, K-feldspar, and biotite) and made an equivalent circuit to simulate the granite's electric and dielectric properties. First, the electric and dielectric properties of the granite sample and the modal mineral composition of the sample were measured. Considering the granite ore is composed of the minerals and the voids, we calculated their values of electric resistance and capacitance and then created the circuit models to simulate the granite rock. For the simulation works, we prepared 10 types of mineral distributions randomly, and an equivalent circuit model was prepared for each mineral distribution. The simulation results of our works showed good agreement with the measured data of *I–V* properties and dielectric properties.

2. Materials and Methods

2.1. Sample

Our method of using an equivalent circuit to simulate rocks can be applied to any type of rock. We selected granite as an example of hard rock for this study. Our sample granite was bought from Kenis, Japan, and the sample was collected from Akaiwa city of Okayama prefecture, Japan. The sample granite was thinly sliced (with a thickness of about 1.1 mm) and its surface was well polished for measurements, including the ones for the elemental compositions as well as the electrical and dielectric properties. Figure 1a shows a snapshot of the granite sample. The elemental mapping was obtained by using a micro-X-ray fluorescence spectrometer (μXRF, M4 Tornedo Plus; Bruker, Billerica, MA, USA) and the mineral composition was then calculated by stoichiometry. Based on the imaging and X-ray analysis, whose results are shown in Figure 1b, the modal composition of the granite sample was calculated. This granite sample consists of quartz (30.1%), plagioclase (37.3%), K-feldspar (23.3%), and biotite (9.2%). The modal composition indicates that the granite is a monzogranite [19]. The modal composition was calculated, considering the composition of the abovementioned four minerals, but only without counting the porosity of the granite sample. In addition, the value of porosity (void% in rock) was needed for our simulation work and we obtained that information from the literature, which will be discussed in Section 2.3.

2.2. Experimental

The electric current–voltage (*I–V*) properties of our granite sample were measured. Granite is a well-known dielectric material [3–6] whose resistivity is quite large (10^{10} $\Omega \cdot$m [20]). Thus, when measuring the electric properties of granite, the surface leakage current should be avoided. The sample was placed in a resistivity chamber (Model 24, ADCMT, Saitama, Japan) where the leakage current of the sample was eliminated by a guard electrode of the chamber. The diameter of the upper electrode of this chamber was 6 mm. An ultra-high resistance meter (5451, ADCMT) was used to measure the *I–V* properties of the sample. In order to measure the dielectric properties of the sample, we used an LCR meter (ZM2376, NF, Yokohama, Japan) with the same resistivity chamber. The capacitance *C* and dielectric loss *tanδ* were measured with the range of frequency from 1 Hz to 1 MHz with 500 measurement points of frequency.

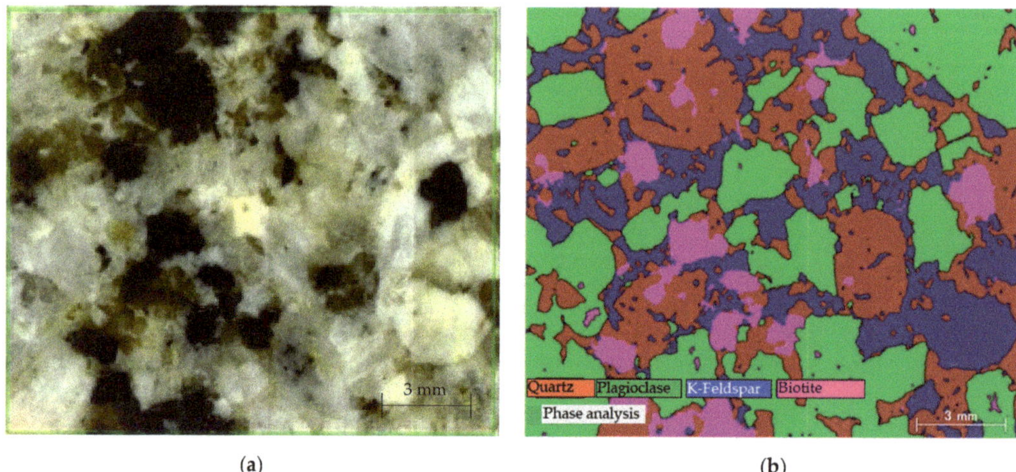

(**a**)   (**b**)

Figure 1. Granite sample. (**a**) Optical microscopic image. (**b**) Micro-X-ray fluorescence spectrometer image identifying quartz (red), plagioclase (green), K-feldspar (blue), and biotite (pink).

2.3. Experimental

The electric current–voltage (*I–V*) properties of our granite sample were measured. Granite is a well-known dielectric material [3–6] whose resistivity is quite large (10^{10} $\Omega \cdot$m [20]). Thus, when measuring the electric properties of granite, the surface leakage current should be avoided. The sample was placed in a resistivity chamber (Model 24, ADCMT, Saitama, Japan) where the leakage current of the sample was eliminated by a guard electrode of the chamber. The diameter of the upper electrode of this chamber was 6 mm. An ultra-high resistance meter (5451, ADCMT) was used to measure the *I–V* properties of the sample. In order to measure the dielectric properties of the sample, we used an LCR meter (ZM2376, NF, Yokohama, Japan) with the same resistivity chamber. The capacitance *C* and dielectric loss *tanδ* were measured with the range of frequency from 1 Hz to 1 MHz with 500 measurement points of frequency.

Any kind of rock contains voids (pores), though their volumes vary with minerals inside the rocks. Voids are not seen in the modal compositions of our granite sample from μXRF mapping, but we can assume their proportions from the literature [11]. Voids' electrical characteristics can be understood by the gas discharge phenomena [13]. Air is one of the dielectric materials whose dielectric breakdown under a HV application is a large branch of interest in HV engineering [13,19]. In order to understand and simulate the behavior of voids in rocks, we also measured the *I–V* properties of the gas discharge by the same experimental procedure mentioned above.

2.4. Simulation Methods

In this study, the equivalent circuit model was used for simulation. An equivalent circuit model can simulate any kind of rock, however, here, we will discuss granite. The granite was assumed to be a cube of 1 mm edge length, and the model was composed of 1000 of 0.1 mm × 0.1 mm × 0.1 mm smaller cubes (Figure 2a). The model was created by assigning the minerals and voids to those smaller cubes of a 0.1mm edge length and making up the granite. The modal mineral composition of the granite sample discussed in Section 2.1 was used for the simulation. The porosity (i.e., volume % of voids) varies depending on the type and origin of the minerals as well as rocks. For granite, the literature reported that the porosity was from 0.9% to 2.6% [21,22]. In our work, the equivalent circuit model was created by assuming that the porosity in the granite was 2 vol%. The electrical properties of the granite were calculated by simulating the electrical properties of the minerals in the granite, and the circuit parameters were calculated for the equivalent circuit models. A circuit-preparing software program written in C# was developed to create a circuit file in the Netlist format, and then the circuit simulations were performed by using the LTspice circuit simulator.

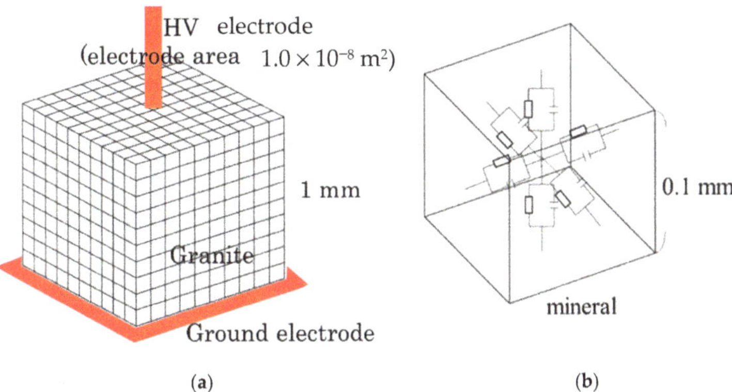

Figure 2. Concept of equivalent circuit model. (**a**) Granite model and (**b**) equivalent circuit model for a small, divided cube of granite.

The possible combinations of mineral distributions can be thousands of patterns. We generated 10 patterns by using random distributions that can decide which minerals or voids will be placed in among the 1000 cubes (0.1 mm edge length). First, void-equivalent circuit elements were placed in the divided cubes with a probability of 2%, and then mineral-equivalent circuit elements for a void were randomly placed in the remaining cubes with the ratio of minerals determined by our μXRF measurement (Figure 1b), based on the fact that the four minerals considered in granite have very similar specific gravities (i.e., quartz (2.65 [11]), plagioclase (2.56 [11]), K-feldspar (2.63 [11]), biotite (2.7–3.4 [23])). Figure 3 shows an example of a mineral distribution using the software of our simulation program. Figure 3a is the 3D model of the minerals and voids (pores) and Figure 3b is the XY plane image of the granite. As mentioned above, the placement of the minerals and voids was selected randomly by our software. When the distribution of the minerals and pores/voids were settled by our software, an equivalent circuit was created automatically to represent the granite sample with the determined distributions of minerals and voids.

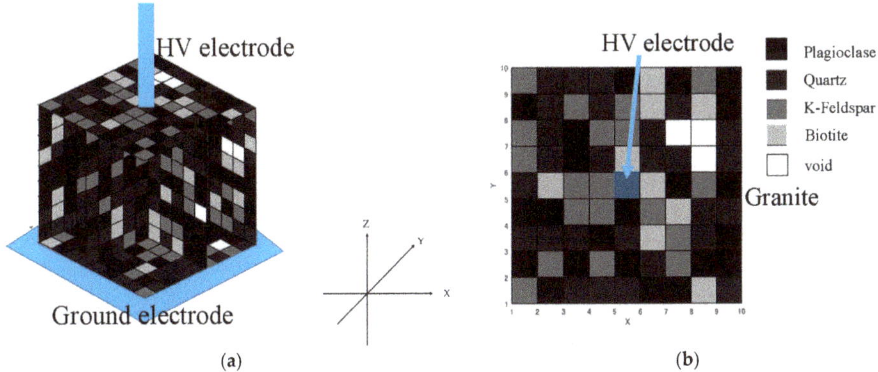

Figure 3. Example of mineral distribution in the simulated granite model. (**a**) Example of a granite 3D model, (**b**) XY plane of the granite model shown in (**a**).

3. Results and Discussion

3.1. Electrical Properties

Following the instructions of the ultra-high resistance meter, we started to take the data of the *I–V* properties from the input voltage of 300 V. The applied voltage was then increased from 300 V to 1000 V with an increment of 100 V. The measurements were carried out with an automatic measurement system developed by the LabVIEW software. The data were taken five times under each voltage and their average values and standard deviations were calculated. The relationship of *J–E* was calculated from the measured data of the *I–V* properties (see Figure 4). Here, *J* represents the electric current density and *E* represents the electric field calculated from the input voltage divided by the distance between the electrodes (i.e., the thickness of the sample, 1.1 mm). *J* was calculated by dividing the electric current *I* with the surface area with the upper electrode, whose diameter was 6 mm. Figure 4 is a double logarithmic graph. The relationship between *J* and *E* can be expressed in the following equation [13]:

$$J = \sigma E \tag{1}$$

where σ is conductivity and the slope of the *J–E* relationship indicates the conductivity σ of the sample. The average value of conductivity was found to be 9.6×10^{-11} S/m and its standard deviation was 9.0×10^{-12} S/m. Thus, the value of resistivity ($1/\sigma$) of our sample was 1.0×10^{10} $\Omega \cdot$m, and it was similar to the value found in the literature (10^{10} $\Omega \cdot$m) [20].

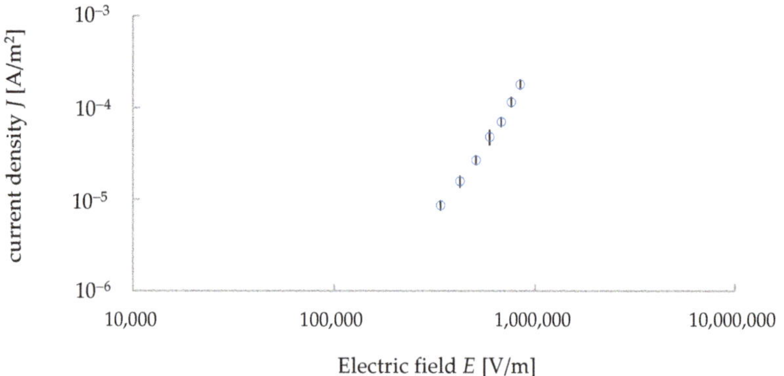

Figure 4. *J–E* relationship of granite sample (measured data) with error bars showing the standard deviation values.

3.2. Dielectric Properties of Granite

The capacitance *C* and dielectric loss *tanδ* were measured with the range of frequency described in Section 2.1 (1 Hz–1 MHz). The measurements were carried out with 500 measurement points over this frequency range. They were measured five times at each frequency and their average and standard deviation values were calculated. The measured data were plotted in double logarithmic graphs (Figure 5). The bars indicate the standard deviation of the values (average data ± standard deviation value). As shown in Figure 5b, the values of *C* decreased with frequency until 1000 Hz but they became almost constant from 1000 to 10,000 Hz. In the latter range, the average value of *C* was 2.45 pF and its standard deviation was 0.253 pF. There were some peak values found in the *C–f* properties. These types of peaks in capacitance *C*, as well as the dielectric constant, were found in the dielectric materials in a different range of frequencies, indicating dielectric relaxation [24]. Dielectric relaxation occurs in composite/polymer materials. Dielectric polarization occurs in dielectric materials when a voltage is applied to them. For an AC source of voltage, the re-orientational motions of dipoles and the translational motions of a charged area differ for different types of materials or polymer chains in a composite.

This difference, seen at the re-orientational motion, appeared in a resonance form with some frequencies and, hence, the dielectric relaxation occurred in the composite/polymer materials. It is to be noted, however, that the imaginary part of the permittivity shows the dielectric relaxation more clearly [24], but as it is not a matter of discussion for this paper, we have avoided the details here. Peak values were also found in $tan\delta$. The average value of $tan\delta$ from the range of 1000 to 10,000 Hz was 0.773 and the average standard deviation was 0.151. The dielectric constant ε was calculated from the Equation (2) [13,25]:

$$C = \varepsilon \frac{S}{L} \tag{2}$$

where C is the capacitance, ε ($\varepsilon = \varepsilon_r\varepsilon_0$) is the dielectric constant, L is the thickness of the sample (1.1 mm) and S is the surface area of the electrode (28.3×10^{-6} m^2). E_r is the relative dielectric constant/permittivity of granite (i.e., 4–7 at 100 MHz) [26,27] and ε_0 is the dielectric constant of vacuum/air (8.85×10^{-12} F/m). From our measurements, the average value of ε_r was found to be 10.8 at the range of frequency from 1000 to 10,000 Hz (10 kHz). The value was not exactly the same as the literature values (i.e., 4–7 at 100 MHz) [26,27]. However, it should be noted that the frequency used in the above reference was 100 MHz, and in general, the capacitance C and the relative dielectric constant ε_r decreased with frequency. Using the calculated values of ε, the ε-f properties of our granite sample were plotted in a double logarithmic graph (Figure 6). The resistance R is expressed by the following equation, which was used to calculate the general resistance value [25]:

$$R = \frac{1}{\sigma}\frac{L}{S} \tag{3}$$

where R is resistance, σ is the electrical conductivity, L is length (i.e., 1.1 mm), and S is the cross-sectional area (i.e., the surface area of the upper electrode, 28.3×10^{-6} m^2). The problem of an electromagnetic wave in dielectric materials (conductivity $\sigma \neq 0$) can be understood by solving Maxwell's equations for electromagnetics [25]. We will not discuss the details here, but by solving Maxwell's equations for dielectric materials, the following relationship can be found [25]:

$$tan\delta = \frac{\sigma}{\omega\varepsilon} \tag{4}$$

where $tan\delta$ is the dielectric loss, ε is the dielectric constant and ω represents the angular frequency (i.e., $\omega = 2\pi f$ where f is frequency). Taking the logarithm of both sides of this equation, we obtain Equation (5) [25],

$$\log tan\delta = \log \sigma - \log \omega - \log \varepsilon \tag{5}$$

This equation reflects the behavior of the dielectric materials under the electromagnetic wave, though it does not reflect the dielectric relaxation phenomenon [25,26]. Due to the polarization in a dielectric material (i.e., granite), the change in the dielectric constant/permittivity depends on the frequency f of the applied electric field. There were some peaks in the dielectric properties (i.e., C: 1.68×10^{-12} F (393 Hz), 8.44×10^{-13} F (21 kHz), 1.59×10^{-12} F (435 kHz), 4.37×10^{-13} F (761 kHz)—Figure 5a; $tan\delta$:0.109 (18.2 kHz), 0.479 (33.4 kHz), 0.0682 (533 kHz)—Figure 5b; ε: 6.54×10^{-11} F/m (393 Hz), 3.28×10^{-11} F/m (21 kHz), 6.59×10^{-11} F/m (435 kHz), 1.70×10^{-11} F/m (761 kHz)—Figure 6) at different frequencies in the measured data of the sample. From Equations (2), (4), and (5), it is clear that the values of capacitance C and dielectric constant/permittivity ε have a negative correlation with the angular frequency ($\omega = 2\pi f$)/frequency f. The slope of $\log \varepsilon$ at the range of frequency 1 to 1000 Hz was found -1.12 (Figure 6). The theoretical value of this slope is -1 (i.e., Equation (5)) and, thus, our measurement results of the dielectric properties of granite show the same trends as the theoretical analysis of dielectric materials. However, after 1000 Hz of frequency, the values of ε started to possess peak values, which indicate the dielectric relaxation in the granite, and as Maxwell's equations on electromagnetics cannot explain the dielectric relaxation of composite materials, our simulation works would not be able to simulate some of the measured data obtained in that frequency range.

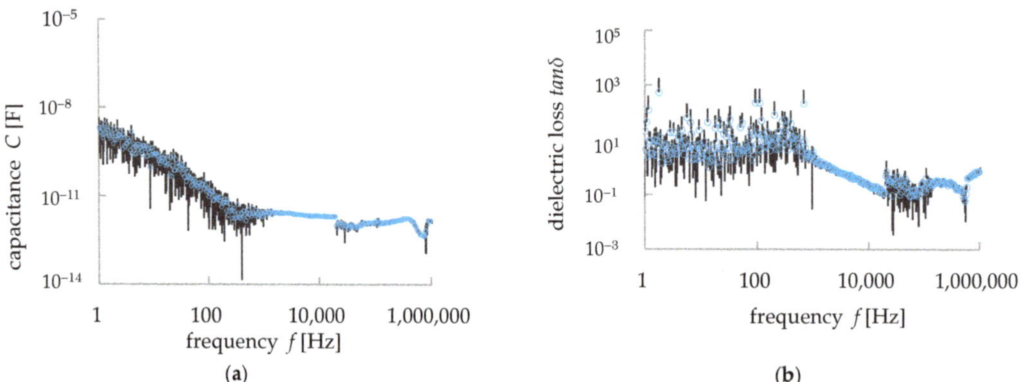

Figure 5. Measured dielectric properties of granite with error bars showing the standard deviation values. (**a**) *C–f* relationship, (**b**) *tanδ–f* relationship.

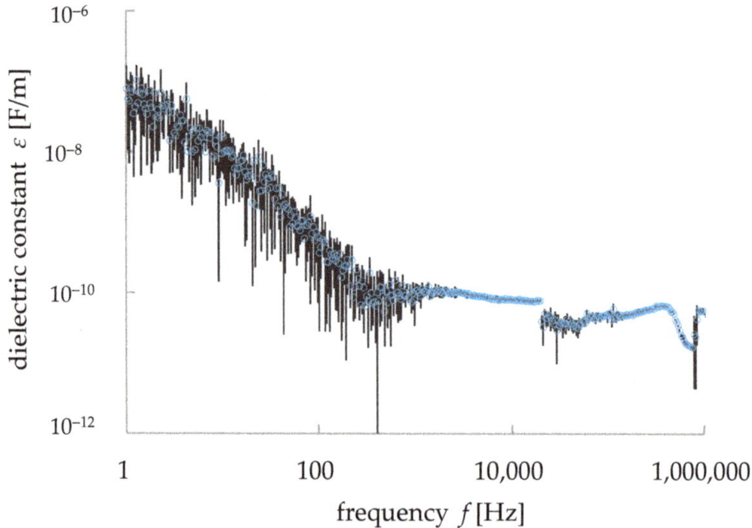

Figure 6. *ε-f* properties of granite calculated from the measured results of dielectric properties with error bars showing the standard deviation values.

3.3. Electrical Properties of a Void (Pore in Mineral/Rock)

Two parallel plate electrodes of aluminum were used to measure the *I–V* properties of the air to understand the electric and dielectric behaviors of voids (pores) in granite. The diameter of the electrodes was 4 mm, and the gap was 0.2 mm for the measurement. An ultra-high resistance meter (R8340A, ADVANTEST, Chiyoda ku, Japan) was used to apply the input voltage and measure the electric current. Figure 7 shows the *I–V* properties of the air. The current at low-applied voltages was almost negligible, about 150 pA from 5 V to 669 V. On the other hand, at 670 V the current rapidly increased to 0.1 mA, and at 680 V the dielectric breakdown occurred. Thus, the dielectric-breakdown voltage of air was found from this measurement at 680 V/0.2 mm (3.4×10^6 V/m). The dielectric-breakdown voltage of air depends on the air pressure, temperature, etc., and in general, it is known at 3.0×10^6 V/m (3.0 kV for 1 cm of gap length), thus, our result of air breakdown was very similar to other reports [13,20]. From our measurement results, it is clear that the conductivity of a void (pore) is a voltage-dependent parameter.

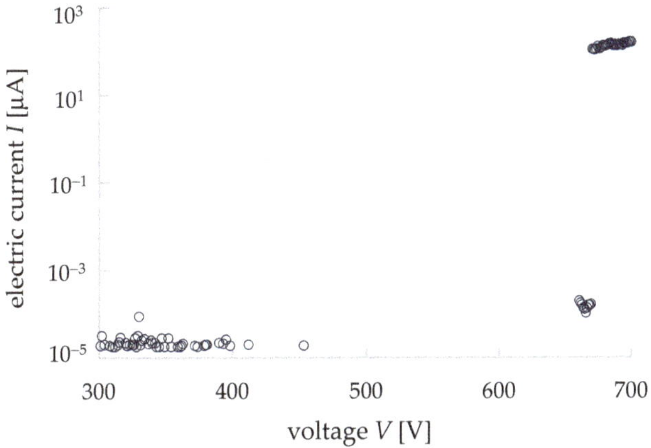

Figure 7. *I–V* properties of the air (i.e., similar to pores or voids in rocks).

3.4. Equivalent Circuit Models for Minerals

The electrical conductivity σ in minerals depends on the temperature. The relationship between the temperature and electrical conductivity σ is expressed by the following equation [11]:

$$\sigma = Aexp\left(\frac{-B}{k_b T}\right) \tag{6}$$

where A and B are constants, T is the absolute temperature, and k_b is Boltzmann's constant (8.618×10^{-5} eV/K). The constants A and B for each mineral in the granite, and the relative permittivity ε_r used to calculate the capacitance are shown in Table 1. In this study, we used an absolute temperature of 293.15 K, which is considered as room temperature (i.e., 20 °C). As shown in Figure 2b, two resistive elements were placed vertically in one cube, and the value of one resistive element was $R/2$. The resistance values R of the minerals were calculated using Equation (3). In this study, the capacitance of a mineral was calculated from Equation (2). Since the value of C is the value of the entire mineral in the cube of Figure 2b, the value of C for the mineral element was $2C$. The values of L and S in Equations (2) and (3) were $L = 1.0 \times 10^{-4}$ m and $S = 1.0 \times 10^{-8}$ m^2 because the minerals were arranged in 0.1 mm cubes, as shown in Figure 2b.

Table 1. Parameters used in this simulation work [9,11].

Mineral	Log(A) [log(s/m)]	B [eV]	ε_r
Quartz	6.3	0.82	6.53
Plagioclase	0.041	0.85	6.91
K-Feldspar	0.11	0.85	6.2
Biotite	−13.8	0	9.28

3.5. Equivalent Circuit Model of Void (Pore in Mineral/Rock)

A void is a pore in a mineral. In this study, we assumed that the void has the same electrical characteristics as the air. Normally, the air does not conduct current when an electric field is applied. However, when a high voltage is applied, a dielectric breakdown can occur, and a very large breakdown-current flows rapidly. The literature indicated that the breakdown voltage of the air is 3 kV for a 1 mm gap of electrodes and the breakdown current varies up to the scale at kA [13]. Therefore, the resistance of a void (pore) can be simulated by a voltage-dependent resistance (VDR), and the capacitor is substituted to create the equivalent circuit model for voids (pores).

In this study, the granite rock was represented by 1000 finite small cubes of 0.1 mm $\times$ 0.1 mm $\times$ 0.1 mm, and the minerals were supposed to be placed in the small cubes. From the electric and dielectric properties of the minerals (i.e., the conductivity and dielectric constant/permittivity, respectively), the circuit parameters were calculated. When a void was present in a cube with a 0.1 mm edge length within the granite rock, the breakdown voltage was 340 V/0.1 mm (=640 V/0.2 mm, Section 3.3), as we determined from our experiment on the air-discharge phenomenon. The value of the capacitor used in the equivalent circuit model for a void was calculated from the size of the cube and the dielectric constant ε of air (i.e., 8.85×10^{-12} F/m). As Equation (2) is applicable for any dielectric materials that include a void (pore), this equation was used to calculate the void capacitance that was 0.885×10^{-15} F and was used in our simulation work. As we described before, the conductivity of air depends on the input voltage, thus, we used a voltage-dependent resistance (VDR) system which can fit with the measurement data of the I–V properties of air. The equivalent circuit was simulated with the LTspice circuit simulator. The I–V properties of air (void) were simulated for the input voltage ranging from 1 to 350 V. The I–E characteristics of the simulation data of a void (pore/air) were prepared and plotted in a double logarithmic graph. The equivalent circuit model of the void and the I–E characteristics of the void are shown in Figure 8a,b, respectively. The electrical properties of a void (see Figure 8) indicates that before the dielectric breakdown of the air, the electric current through a void can be negligible as the value of the electric current is very low (i.e., 10^{-12} A at 1 to 350 V) when comparing to that of near the dielectric-breakdown area (i.e., 10^{-3} A at 680 V) of a void. We aimed to make an equivalent circuit model for the void, which can represent the I–V properties in a way that could fit the dielectric-breakdown region of the void. Figure 8b compared the I–E properties of the void obtained via measurement and simulation. From the simulation results, the electric current was 20 pA for the input voltage of 10 V (i.e., 100,000 V/m), and it remained less than 10^{-9} A even at 150 V (i.e., 1,500,000 V/m). The values of the electric current were 0.1×10^{-9} at 50 V (i.e., 500,000 V/m) and 0.21×10^{-9} A at 100 V (1,000,000 V/m)). The electric current increased gradually with the increase in input voltage, as we found it was 15.7×10^{-9} A at 200 V (i.e., 2,000,000 V/m) and 0.172×10^{-6} A at 300 V (i.e., 3,000,000 V/m). Finally, the current reached 1 mA in between the input voltage of 340 V (3,400,000 V/m) and 350 V (i.e., 35,000,000 V/m). The dielectric-breakdown voltage was found in our measurement at 680 V/0.2 mm (i.e., 340 V/0.1 mm), and, thus, we confirmed the reproduction of the I–V characteristics of a void (air) in our simulation work. It is to be noted that, while measuring the dielectric breakdown of the void, the system was very unstable to measure, but from the simulation works, the details of the electric current were calculated.

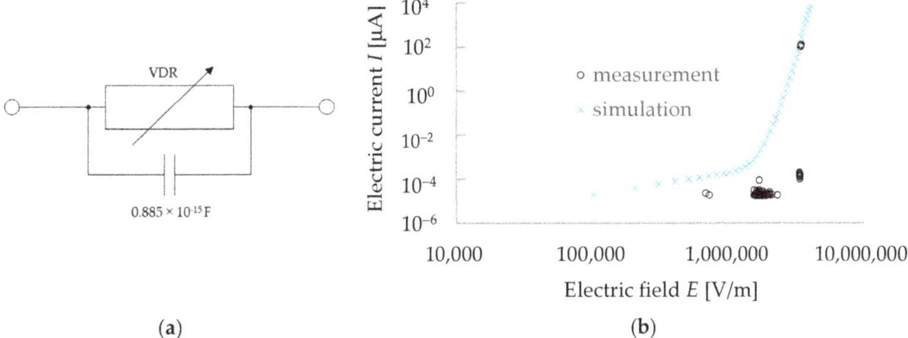

(**a**) (**b**)

Figure 8. Equivalent circuit model and electrical properties of a void (pore) in minerals. (**a**) Equivalent circuit and (**b**) *I–E* properties of a void.

3.6. Simulation Results of Granite Sample

The patterns of minerals and void distributions can be more than thousands, but to evaluate our simulation works, we created 10 distribution patterns of minerals and voids, and then 10 equivalent circuits of granite samples. In order to evaluate the simulation works, a Ltspice circuit simulator was used to simulate the *I–V* properties of the circuits (i.e., granite) under direct-current (DC) input voltages from 100 to 1000 V. They were also simulated with AC voltage to evaluate the dielectric properties of our simulation process with a vast range of frequencies (i.e., 1 Hz to 1000 kHz).

At first, we will discuss the DC properties of our simulation results. The electric current was calculated for each input voltage for each equivalent circuit (i.e., each mineral distribution in granite). From the 10 simulation results of the circuits, the mean values and standard deviation values of the current were calculated for each input voltage. The average *J–E* properties were calculated and plotted in a double logarithmic graph, as seen in Figure 9. The bars indicate the standard deviation of the values (average data ± standard deviation value). We also calculated the conductivity σ of the granite sample for both the measurement and simulation results (Figure 10a). The average values of σ [pS/m] was 53.5 (measurement) and 36.2 (simulation). The standard deviation of σ [pS/m] was 38.5 (measurement) and 2.77 (simulation). The literature value of the resistivity (i.e., the reciprocal of conductivity) of granite was in the order of 10^{12} $\Omega \cdot$cm (i.e., 100 pS/m in the value of conductivity) [21] which is the same as our measured and simulation values. Thus, our simulation results are considered reliable. However, a difference in the conductivity properties of granite was found between the measurement results and simulation works. As shown in Figure 11a, it is clear that the simulation values are almost constant regardless of the input voltages, while the measured values increase with the input voltage. This difference might occur from the Schottky effect observed under a HV applied voltage on dielectric materials [13]. For an electric field *E* at temperature *T*, the emission of electrons/ions can be expressed by the following equation:

$$I = SDT^2 exp\left(\frac{-e}{k_b T}\right) exp\left\{\frac{e}{k_b T}\left(\frac{eE}{4\pi\varepsilon}\right)^{\frac{1}{2}}\right\} \tag{7}$$

where, *D* is the Dushman constant (i.e., 1.20×10^6 A/m² K²), *T* is the absolute temperature in K, φ is the activation energy (i.e., similar to a work function in metal), *e* is the elementary charge quantity, and the others were described before. The details will not be discussed here, however, there is a linear relationship between the natural logarithm of electric current *I* and electric field $E^{1/2}$ (Equations (8)–(10)).

$$LnI = K_0 E^{\frac{1}{2}} + \ln I_0 \tag{8}$$

$$K_0 = (exp/k_bT)(exp/4\pi\varepsilon)^{\frac{1}{2}} \tag{9}$$

$$\ln I_0 = \ln\left(SDT^2\right) - (e\phi) \tag{10}$$

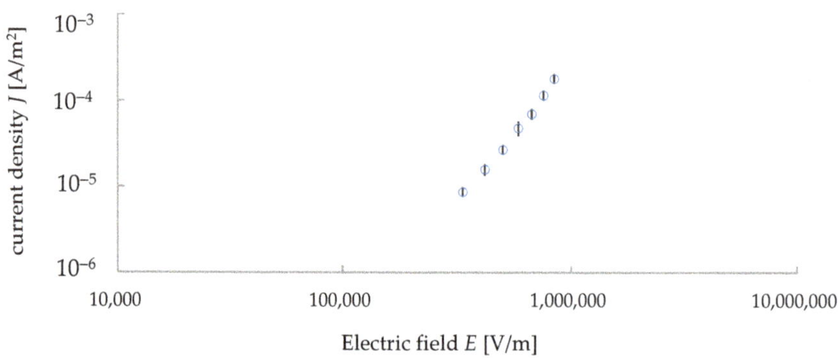

Figure 9. Average of simulation results of *J–E* for granite with error bars showing the standard deviation values.

These relationships can be seen in Figure 10b. For the measured data, the determination coefficient was 0.9991, while the same coefficient was 0.9583 for the simulation results. It is worth noting that the Schottky effect was not considered in this simulation work. The electrical properties of the minerals did not consider any voltage-dependent resistance (VDR), but rather, they used constant electrical and dielectric properties to create an equivalent circuit. It explains why the conductivity did not change with the voltage (Figure 10a). As we used voltage-dependent resistance (VDR) for the voids (pores) that account for only 2% in granite, the voltage change had a very limited effect on the conductivity in the simulated results. However, the mean and standard deviation values of the conductivity were very close to the measured values [21]. In fact, it can be understood from the Schottky equation that the resistivity of the rock changes greatly with the temperature, suggesting that it should be considered in future simulations. Even the recognition of voids (i.e., pores in minerals) and considering them into the equivalent circuit accurately may simulate the HV effect on the rocks more concretely. The other paper by our group discussed this relationship (the *I–V* as a function of temperature) under the HV impulse application on granite [19].

Figure 11 shows the mean values of the capacitance *C* and dielectric loss *tanσ* calculated from our simulation works. The error bars indicate the standard deviation of the values (average data ± standard deviation value). In contrast to the measured values (i.e., Figure 3a), the simulation results showed little frequency dependence. The average value of capacitance *C* was 0.0074 pF in between the frequency region of 1 kHz and 10 kHz. The mean value of the standard deviation in this frequency range was 0.000312 pF. These values are about one-tenth of the measured values (i.e., Figure 5a). The *C–f* characteristics of the simulation results were different from the measured values, and the values were constant even in the low-frequency range. However, the standard deviation showed a large variation up to 1 kHz, and then it became completely constant. This result suggests that the trend of the dielectric properties could be reproduced, although the average value of *C* could not be completely reproduced. In contrast to the measured values, the simulation results for the *tanδ–f* relationship also showed little frequency dependence. The mean value of the loss factor *tanδ* was 0.079, and the mean value of the standard deviation was 0.032. These values were also about one-tenth of the measured values (i.e., Figure 5b). Figure 5 shows that a dielectric relaxation phenomenon occurs around 1 kHz. In the measured data, the peak values for *C* and *tanδ* were found in frequencies other than 1 kHz, but in the simulation results, it was not observed at frequencies other than 1 kHz. However, we can confirm that the variation becomes larger in the low-frequency band below 100 Hz. This may correlate with the several peaks that occurred in the measured values. From this result, the dielectric phenomenon was considered to be reproduced, although the average value of *tanδ* could not be reproduced completely. It is worth noting that, unlike the measured data, this simulation work did not consider the dielectric relaxation phenomenon in the consideration, thus, the values of the simulated dielectric properties (i.e., *C*, *tanδ*) differed from the measured data. On the other hand, the general trend of the dielectric properties was similar between the simulation and measurement data.

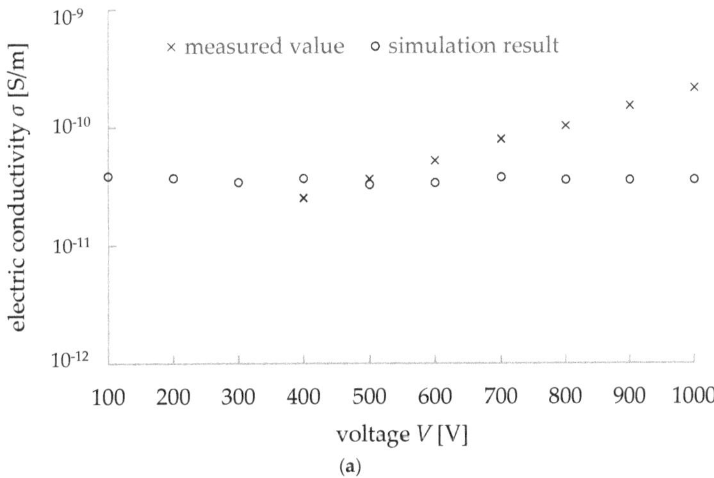

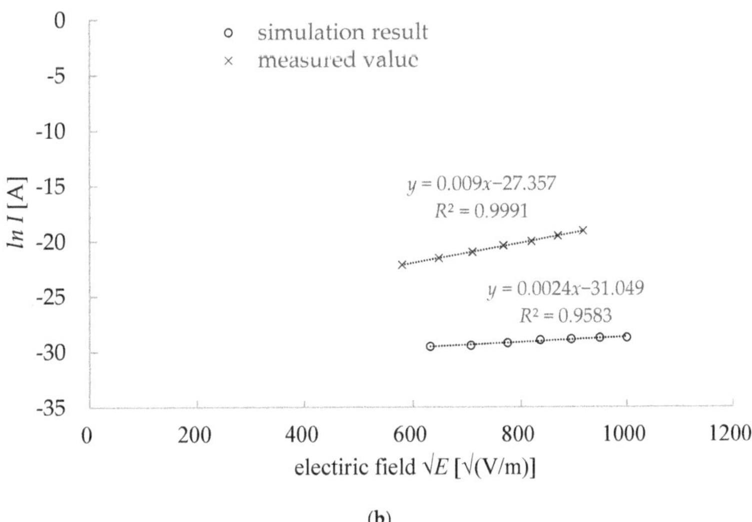

Figure 10. Simulation results and analysis (**a**) simulation results of conductivity of granite, (**b**) Schottky effect on granite at room temperature.

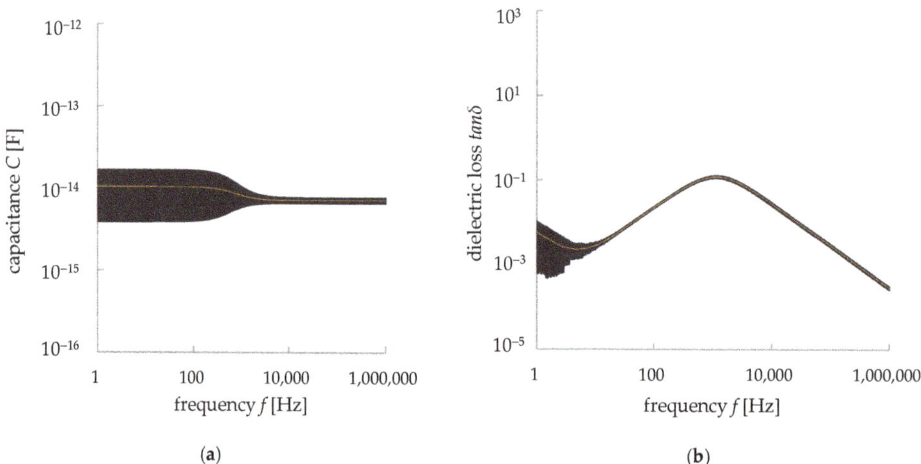

Figure 11. Simulation results of dielectric properties with error bars showing the standard deviation values. (**a**) $C–f$ relationship, (**b**) $tan\delta–f$ relationship.

The dielectric constant was calculated from the simulation results of each of the 10 patterns of equivalent circuit simulation, and the average value was calculated. Figure 12 shows the comparison between the calculated results and the measured data. As shown in Figure 12, the dielectric relaxation occurred in the measured dielectric constant values but was hardly observed in the simulation results. However, the measured and simulated dielectric constant values were close to each other in their magnitude. In addition, in the range of frequency between 1 Hz to 1 kHz, the measured values showed a straight line with a negative slope and a constant value after 1 kHz, while the simulation results showed a slight decrease and a constant value after 1 kHz. Thus, it is clear that, considering Maxwell's equation for the dielectric materials (i.e., most of the rocks and granite studied in this work), the dielectric behavior for a certain range of frequency (1–100 Hz for granite in this work) can be reproduced by our simulation method to some extent.

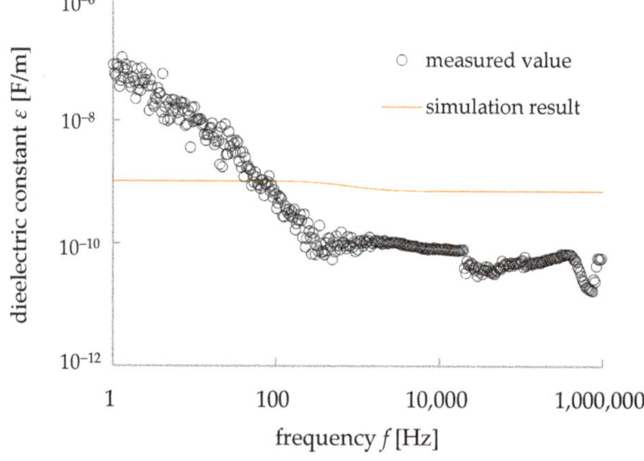

Figure 12. Comparison of dielectric constant with frequency in between the measurement and simulation results.

However, it is to be noted that the equivalent circuit model method has some limitations, as recognized by our simulation results of the dielectric properties and dielectric relaxations. Capacitance values, as well as dielectric parameter/permittivity, are strongly time-dependent (i.e., frequency) parameters under a certain applied voltage. However, the capacitance value of a capacitor in the equivalent circuit model is fixed, as the parameter of capacitance cannot be replaced with a time-dependent parameter in circuit simulator software. Dielectric relaxations can be observed in composite materials with different frequencies of AC properties and are also observed in our dielectric measurements of granite. With our current model, the simulation results could not imitate the dielectric relaxations. However, as the parameter $tan\delta$ is determined with both dielectric and electric properties (i.e., Equations (4) and (5)), a part of the dielectric relaxation properties of $tan\delta$ was reproduced by our results. These basic disadvantages can be solved by considering several granite samples with a more accurate knowledge of the compositions and their dielectric properties. Further experiments will be needed to improve this method, which will be the next step in our research interests.

4. Conclusions

In this study, equivalent circuit models of granite were developed by considering the distributions of minerals and voids in granites. In order to confirm the validity of the equivalent circuit model, we measured the electrical and dielectric characteristics of the granite sample and compared them with the simulation results. The results presented in this article are summarized as follows:

(1)   The calculated electrical conductivity of the granite model and the actual granite were close to each other (measurement: 53.5 pS/m, simulation: 36.2 pS/m), and the standard deviation was very small in the simulation results (i.e., 2.77 pS/m).

(2)   The Schottky effect was observed in the I–V properties of granite, and it emphasized the necessity to consider the dielectric constant in order to simulate the I–V properties.

(3)   Comparing the simulation values of C and $tan\delta$ of the granite model and the measurement data, the dielectric relaxation phenomenon was observed in both the simulation and measurement data for the $tan\delta$ and frequency (f) relationship, and their values were close to each other. However, our simulation works did not imitate the relaxation phenomenon for the C–f relationship. For the higher frequencies (i.e., larger than 1 kHz), the simulation results showed a larger dielectric constant ε (i.e., 10^{-9} F/m), while the measured values were around 10^{-10} F/m.

These results suggested that the simulation model provided electrical properties similar to those of the granite sample experimentally measured. We prepared equivalent circuit models to simulate granite, but the models can be applied to any rock in order to understand the electrical and dielectric phenomena in rocks. A model considering both conductivity and the dielectric constant will be useful for understanding the electric pulse comminution. Further research works will be carried out to reproduce the dielectric properties of granite with the equivalent circuit model. We are working on a more equivalent circuit method for different ores to develop this method to be more effective in understanding the electric-pulse application behaviors of different minerals, as well as ores.

Author Contributions: Conceptualization, M.K. and A.O.; methodology, K.F., M.K., K.K., N.O., M.F. and A.O.; software, K.F., M.K. and K.K.; validation, K.F., M.K., K.K. and A.O.; formal analysis, K.F., M.K. and K.K.; investigation, K.F., M.K. and K.K.; resources, M.K. and A.O.; data curation, K.F., M.K., K.K. and A.O.; writing—original draft preparation, K.F. and M.K.; writing—review and editing, K.F., M.K. and A.O.; visualization, K.F., M.K. and K.K.; supervision, M.K. and A.O.; project administration, M.K. and A.O.; funding acquisition, M.K. and A.O. All authors have read and agreed to the published version of the manuscript.

Funding: This research received no external funding.

Institutional Review Board Statement: Not applicable.

Informed Consent Statement: Not applicable.

Data Availability Statement: Not applicable.

Acknowledgments: We acknowledge Manabu Mizuhira of Bruker Japan for the elemental map analyses of the sample granite at their application laboratory.

Conflicts of Interest: The authors declare no conflict of interest.

References

1. Andres, U.; Timoshkin, I.; Jirestig, J.; Stallknecht, H. Liberation of valuable inclusions in ores and slags by electrical pulses. *Powder Technol.* **2001**, *114*, 40–50. [CrossRef]
2. Tester, J.W.; Anderson, B.J.; Batchelor, A.S. Impact of enhanced geothermal systems on US energy supply in the twenty -first century. *Phil Trans. Math. Phys. Eng. Sci.* **2007**, *365*, 1057–1094. [CrossRef] [PubMed]
3. Sasaki, Y.; Fujii, A.; Nakamura, Y. Electromagnetic Parameters of Rocks in Short Wave Band—Influences of Lithologic Characters. *J. Jpn. Soc. Eng. Geol.* **1993**, *34*, 109–119. [CrossRef]
4. Yokoyama, H. Dielectiric Constant of Rocks in the Frequency Range 30 Hz to 1 MHz—Studies on the dielectric properties of rocks(1). *J. Min. Inst. Jpn.* **1997**, *93*, 347–352. [CrossRef]
5. Liberation of Minerals with Electric Pulse (Digital Description Of Prof. Syuji Owada's Work, Waseda University, Japan). Available online: https://www.waseda.jp/top/news/6137 (accessed on 20 December 2021). (In Japanese).
6. Takahashi, H.; Ishihara, H.; Umetani, T.; Shinohara, T. Development of Prediction Methods for Temperature Distribution and Fractured Volume in Granite Under Microwave Radiation. *Tech. Note Port. Harb. Res. Inst. Minist. Transp. Jpn.* **1986**, *558*, 3–20, (In Japanese but Synopsis written in English).
7. Wang, E.; Shi, F.; Manlapig, E. Factors affecting electrical comminution performance. *Miner. Eng.* **2012**, *34*, 48–54. [CrossRef]
8. Seyed, M.R.; Bahram, R.; Mehdi, I.; Mohammad, H.R. Numerical simulation of high voltage electric pulse comminution of phosphate ore. *Int. J. Min. Sci. Technol.* **2015**, *25*, 473–478.
9. Li, C.; Duan, L.; Tan, S.; Chikhotkin, V. Influences on High—Voltage Electro Pulse Boring in Granite. *Energies* **2018**, *11*, 2461. [CrossRef]
10. Zuo, W.; Shi, F.; Manlapig, E. Modelling of high voltage pulse breakage of ores. *Miner. Eng.* **2015**, *83*, 168–174. [CrossRef]
11. Walsh, S.D.C.; Lomov, I.N. Micromechanical modeling of thermal spallation in granitic rock. *Int. J. Heat Mass Tran.* **2013**, *65*, 366–373. [CrossRef]
12. Walsh, S.D.C. *Modeling Thermally Induced Failure of Brittle Geomaterials*; Technical Report; Office of Scientific & Technical Information Technical Reports, USA; 2013; Available online: https://digital.library.unt.edu/ark:/67531/metadc828604/ (accessed on 11 January 2022).
13. Küchler, A. *High Voltage Engineering*; Springer Vieweg: Berlin, Germany, 2018.
14. Zuo, Z.; Dissado, L.A.; Yao, C.; Chalashkanov, N.M.; Dodd, S.J.; Gao, Y. Modeling for Life Estimation of HVDC Cable Insulation Based on Small-Size Specimens. *IEEE Electr. Insul. Mag.* **2020**, *36*, 19–29. [CrossRef]
15. Kabir, M.; Suzuki, M.; Yoshimura, N. Influence of non-uniform grain distribution of grains on noise absorption characteristics of ZnO varistors. *IEEJ Trans. Fundam. Mater.* **2007**, *127*, 35–40. [CrossRef]
16. Ono, Y.; Kabir, M.; Suzuki, M.; Yoshimura, N. Analysis of Electrical Properties of Insulation Material Filled with ZnO Microvaristors. *Electr. Eng. Japan* **2013**, *133*, 882–887. [CrossRef]
17. Kabir, M.; Suzuki, M.; Yoshimura, N. Analysis of ZnO Microvaristors with Equivalent Circuit Simulation. In Proceedings of the 2011 Electrical Insulation Conference (EIC2011), Annapolis, MD, USA, 5–8 June 2022.
18. Fukushima, K.; Kabir, M.; Kanda, K.; Obara, N.; Fukuyama, M.; Otsuki, A. Simulation of Electrical and Thermal Properties of Granite under the Application of Electrical Pulse Using Equivalent Circuit Models. *Materials* **2022**, *15*, 1039. [CrossRef] [PubMed]
19. Le Bas, M.J.; Streckeisen, A.L. The IUGS systematics of igneous rocks. *J. Geol. Soc. Lond.* **1991**, *148*, 825–833. [CrossRef]
20. Rocks: Electrical Properties. Available online: https://www.britannica.com/science/rock-geology/Electrical-properties (accessed on 11 January 2022).
21. Hosokawa, T.; Yumoto, M. Discharge Phenomena in Vacuum: Discharge Mechanisms and Its Applications. *IEEJ Trans. Fundam. Mater.* **1994**, *114*, 77–83. [CrossRef]
22. Duba, A.; Piwinskii, A.J.; Santor, M.; Weed, H.C. The electrical conductivity of sandstone, limestone and granite. *Geophys. J. R. Astr. Soc.* **1978**, *53*, 583–597. [CrossRef]
23. Minerals Silicates Minerals: Biotite. Available online: https://geologyscience.com/minerals/biotite/ (accessed on 28 March 2022).
24. Cole, K.S.; Cole, R.H. Dispersion and Absorption in Dielectrics I. Alternating Current Characteristics. *J. Chem. Phys.* **1941**, *9*, 341–351. [CrossRef]
25. Sadiku, M.N.O. *Elements of Electromagnetics*; Oxford University Press: Oxford, UK, 2011.
26. Sueyoshi, K.; Yokoyama, T.; Katayama, I. Experimental Measurement of the Transport Flow Path Aperture in Thermally Cracked Granite and the Relationship between Pore Structure and Permeability. *Hindawi Geofluids* **2020**, *2020*, 8818293. [CrossRef]
27. Davis, J.L.; Annan, A.P. Ground-penetrating Radar for High-Resolution Mapping of Soil and Rock Stratigraphy. *Geophys. Prospect.* **1989**, *37*, 531–551. [CrossRef]

Article

Non-Destructive Imaging on Synthesised Nanoparticles

Kelvin Elphick [1], **Akinobu Yamaguchi** [2], **Akira Otsuki** [3,4,5], **Neil Lonio Hayagan** [3] and **Atsufumi Hirohata** [1,*]

1   Department of Electronic Engineering, University of York, Heslington, York YO10 5DD, UK;
    kelvin.elphick@gmail.com
2   Laboratory of Advance Science and Technology for Industry, University of Hyogo, Hyogo 678-1205, Japan;
    yamaguti@lasti.u-hyogo.ac.jp
3   Ecole Nationale Supérieure de Géologie, GeoRessources UMR 7359 CNRS, University of Lorraine, 2 Rue du
    Doyen Marcel Roubault, BP 10162, 54505 Vandoeuvre-lès-Nancy, France; akira.otsuki@univ-lorraine.fr (A.O.);
    hayaganneil@gmail.com (N.L.H.)
4   Waste Science & Technology, Luleå University of Technology, SE 971 87 Luleå, Sweden
5   Neutron Science Laboratory, The Institute for Solid State Physics, The University of Tokyo,
    Chiba 277-8581, Japan
*   Correspondence: atsufumi.hirohata@york.ac.uk

Abstract: Our recently developed non-destructive imaging technique was applied for the characterisation of nanoparticles synthesised by X-ray radiolysis and the sol-gel method. The interfacial conditions between the nanoparticles and the substrates were observed by subtracting images taken by scanning electron microscopy at controlled electron acceleration voltages to allow backscattered electrons to be generated predominantly below and above the interfaces. The interfacial adhesion was found to be dependent on the solution pH used for the particle synthesis or particle suspension preparation, proving the change in the particle formation/deposition processes with pH as anticipated and agreed with the prediction based on the Derjaguin–Landau–Verwey–Overbeek (DLVO) theory. We found that our imaging technique was useful for the characterisation of interfaces hidden by nanoparticles to reveal the formation/deposition mechanism and can be extended to the other types of interfaces.

Keywords: scanning electron microscopy; backscattered electrons; electron flight simulation; nanoparticles; synthesis

Citation: Elphick, K.; Yamaguchi, A.; Otsuki, A.; Hayagan, N.L.; Hirohata, A. Non-Destructive Imaging on Synthesised Nanoparticles. *Materials* **2021**, *14*, 613. https://doi.org/10.3390/ma14030613

Academic Editor: Miguel Monge
Received: 18 December 2020
Accepted: 25 January 2021
Published: 29 January 2021

Publisher's Note: MDPI stays neutral with regard to jurisdictional claims in published maps and institutional affiliations.

Copyright: © 2021 by the authors. Licensee MDPI, Basel, Switzerland. This article is an open access article distributed under the terms and conditions of the Creative Commons Attribution (CC BY) license (https://creativecommons.org/licenses/by/4.0/).

1. Introduction

Nanoparticles have been synthesised on metallic electrodes and (non-)conductive substrates. Their properties are known to be controlled by their interfacial structures governed by their formation processes. To date, these interfaces have been predominantly imaged by destructive methods, which can achieve nanometric resolution. As reported earlier [1], the highest resolution can be achieved by (scanning) transmission electron microscopy ((S)TEM) and atom probe imaging. These methods have been used commonly for the nano- to atomic-scale analysis of the junction interfaces. However, they require samples to be milled for electron transparency, introducing possible strain and defects during the sample preparation and hindering the direct correlations between the interfacial structures and electromagnetic properties. Electron beam-induced and -absorbed currents (EBIC and EBAC, respectively) has also been used, especially in semiconductor industries, but they are limited to transport properties with the most conductive layer with a sub-micron resolution.

On the other hand, our recently developed non-destructive imaging method can be performed by controlling the acceleration voltage in scanning electron microscopy (SEM) without modifying a sample and a device [2]. This method achieves a high in-plane resolution of a few nm without any additional requirements of sample preparation for imaging. By controlling the electron acceleration voltages in SEM, the penetration depth of the electron beam can be manipulated. The corresponding generation of secondary electrons (SEs) and backscattered electrons (BSEs) are generated within the electron plume introduced. Since SEs can be surface sensitive via following scattering processes within the specimen, BSEs are detected in this non-destructive imaging method using an energy filter. Recently, we have demonstrated in situ imaging capability under the current-voltage applications, allowing direct comparisons with the defects and the electrical transport properties [3]. Further, the combinations of spectroscopic and scattering/reflective chemical analysis allowed us to evaluate the origins of the defects, which is ideal as a quality assurance for nano-electronic industries. The defect details and the corresponding transport properties can be fed back to the processes of the device fabrication processes, improving the yields [4].

In this study, we applied our method to the characterisation of nanoparticles. We prepared two types of nanoparticles by X-ray radiolysis and the sol-gel method. By imaging these nanoparticles using our method, clear differences in their interfacial structures were found. They revealed the differences in their formation processes during the synthesis or particle suspension preparation, and confirmed the formation/dispersion models predicted depending on the solution pH used for the particle synthesis or particle suspension preparation. Hence, our imaging method can be highly useful for the understanding of the particle synthesis/dispersion processes and can be fed back to the process optimisation of nanoparticle systems.

2. Synthesis of Nanoparticles and Preparation of Nanoparticle Suspensions

2.1. X-ray Radiolysis

X-ray radiolysis was used to synthesise nanoparticles using beam line BL8S2 at the Aichi Synchrotron Radiation Center, Aichi Science & Technology Foundation. As detailed in our previous publication [5], a 100-mL aliquot of 0.37 mol/L (M) $Cu(COOCH_3)_2$ (FUJI-FILM Wako Pure Chemical Corporation, Osaka, Wako 1st Grade, Japan) was prepared by diluting the stock solution and mixed with methanol with the volume ratio shown in Table 1. In total, 20 µL of these solutions were spread on Si/SiO_2, n-Si, or Cu substrate, followed by the exposure of 5 min of synchrotron X-ray radiation to synthesise nanoparticles. X-ray irradiation can generate radicals from the radiolysis of liquids and secondary electron generation from substrates dipped in metallic liquid solution. There are some possible routes for the nucleation, ripping, growth, aggregation, and immobilisation of the particles onto the surface of substrate. In particular, near the substrate surface and the interface between the particles and the substrate, the nucleation, growth, and aggregation of these particles can be controlled by the X-ray irradiation significantly. Therefore, this investigation by non-destructive imaging is significantly worthwhile for the understanding of the physical and chemical mechanisms for the synthesis of particles. Simultaneously, this study can also provide the clue to control the synthesis and immobilisation of the particles.

Table 1. List of nanoparticles synthesised by X-ray radiolysis.

Samples	$Cu(COOCH_3)_2$ Volume	Additive Solution and Volume	Substrates
#1	200 µL	Methanol 1 µL	Si/SiO_2
#2	200 µL	Methanol 1 µL	n-Si(001)
#3	200 µL	Methanol 1 µL	p-Si(001)
#4	200 µL	Methanol 1 µL	n-Si(111)
#5	200 µL	Methanol 1 µL	Ni
#6	200 µL	Methanol 1 µL	Al
#7	200 µL	Methanol 1 µL	$LiNbO_3$
#8	200 µL	Methanol 1 µL	PTFE
#9	900 µL	Methanol 5 µL + NH_3 100 µL (pH = 8)	Si/SiO_2
#10	900 µL	Methanol 5 µL + NH_3 200 µL (pH = 9)	Si/SiO_2
#11	950 µL	Methanol 5 µL + NH_3 50 µL (pH = 7)	Si/SiO_2
#12	500 µL	Methanol 5 µL + NH_3 500 µL (pH = 11)	Si/SiO_2

2.2. Nanoparticle Synthesis by the Sol-Gel Method and Suspension Preparation

Monodispersed silica nanoparticles were prepared by using the method proposed by [6]. The average particle radius measured by using TEM images was 280 nm and used in the DLVO (Derjaguin–Landau–Verwey–Overbeek) potential calculation (see Section 3). The stock solution containing synthesised silica particles was washed several times to minimise the salt concentration prior to preparation of the desired silica particle suspensions for investigation with the desired chemical environments. Under the different conditions listed in Table 2, silica particle suspensions were prepared in aqueous salt solution (1×10^{-2} M KNO_3, Sigma-Aldrich (St. Louis, MO, USA), and their pH was adjusted using HNO_3 or KOH followed by conditioning the suspensions for 30 min. A tiny volume of each sample was pipetted and deposited on a standard SEM aluminium stub that was left in an oven at 50 °C for several hours to let the moisture content evaporate and firmly deposit particles on the stub by capillary forces with the residual moisture, followed by metallization of the stub for the sample conductivity.

Table 2. List of the silica nanoparticle suspensions studied.

Samples	Preparation Methods
Tug 2	Silica 0.1 vol.%, 1×10^{-2} M KNO_3, pH10
Tug 3	Silica 0.01 vol.%, 1×10^{-2} M KNO_3, pH10
Tug 5	Silica 0.1 vol.%, 1×10^{-2} M KNO_3, pH2
Tug 6	Silica 0.01 vol.%, 1×10^{-2} M KNO_3, pH2

3. DLVO Potential Calculation

Potential energy calculation between (a) two silica particles or (b) aluminium stub/plate and a silica particle was performed using the DLVO (Derjaguin–Landau–Verwey–Overbeek) theory, which is a well-known theory for describing the material interactions with the summation of the van der Waals potential (V_A) and electrical double layer potential (V_R) [7,8]. If the total potential energy ($V_T = V_A + V_R$) is high and positive, particles repel each other; otherwise, particles attract each other. This is a straight-forward theory, which can explain particle coagulation/dispersion in many different colloidal systems, e.g., [9–15]. In our previous study that is relevant to the present study, the DLVO theory was also applied to investigate the particle–particle interactions in the system with the small quantity of water present in agglomeration processes [13]. The following paragraphs will introduce and explain the equations used for the potential energy calculation.

Equations used to calculate the potential energies between similar spherical particles [7,8]:

$$V_A = -\frac{Aa}{12H} \tag{1}$$

$$V_R = \frac{64\pi a n k T \gamma^2 \exp(-\kappa H)]}{\kappa^2} \tag{2}$$

$$n = N_A C \tag{3}$$

$$\kappa = \left(\frac{8\pi n z^2 e^2}{\varepsilon \varepsilon_0 k T}\right)^{\frac{1}{2}} \tag{4}$$

$$\gamma = \frac{\exp\left(\frac{ze\zeta}{2kT}\right) - 1}{\exp\left(\frac{ze\zeta}{2kT}\right) + 1} \tag{5}$$

where A is the Hamaker (J) constant, a is the particle radius (nm), H is the inter-particle separation distance, n is the number concentration of ions (nm^{-3}) defined in Equation (3), N_A is the Avogadro's number (6.022×10^{23} mol^{-1}), C is the concentration of ions (mol/nm^3), k is the Boltzmann constant (1.38×10^{-23} J/K), T is the absolute temperature (K), γ is the reduced surface potential (unitless), κ is the Debye–Huckel reciprocal length (nm^{-1}) defined in Equation (4), ε is the dielectric constant of the medium, ε_0 is the permittivity of free space (C/Vnm), z is the ionic valence, e is the elementary charge (C), and ζ is the zeta potential (V). The zeta potential values of silica particles [16] and aluminium plate [17] were extracted from the literature and used for the present calculation.

Equations used to calculate the potential energy between plate and spherical particle interactions [18] (in our case, the interaction between the aluminium stub and silica particle):

$$V_A = -\frac{A}{6}\left(\frac{a}{H} + \frac{a}{H + 2a} + \ln\left(\frac{a}{H + 2a}\right)\right) \tag{6}$$

$$V_R = \frac{128\pi ankT\gamma_s\gamma_p \exp(-\kappa H)]}{\kappa^2} \tag{7}$$

where γ_s and γ_p are the reduced surface potential of the sphere and plate (unitless), respectively.

In this article, the calculated total potential energies were normalized by the thermal fluctuation energy (kT). For the dissimilar plate-particle systems, the Hamaker constant A_{132} was calculated by using the following Equation [19]:

$$A_{132} = \left(\sqrt{A_{11}} - \sqrt{A_{33}}\right)\left(\sqrt{A_{22}} - \sqrt{A_{33}}\right) \tag{8}$$

where A_{11} is the Hamaker constant of particle 1 in vacuum, A_{22} is the Hamaker constant of particle 2 in vacuum, and A_{33} is the Hamaker constant of water in vacuum. These values were obtained from the literature [19,20].

4. Non-Destructive Imaging

As described in Section 1, the acceleration voltage of the electron beam in SEM was precisely controlled to achieve the corresponding penetration into the layer above and below the buried interface to be investigated. The detailed procedures of the non-destructive imaging we recently developed can be found in [3]. An electron flight simulator, CASINO [21], was used to calculate the number of BSEs to be generated in nanoparticles. For the cupric and silica nanoparticles investigated in this study, the simulations show that BSEs can be generated in the vicinity of the nanoparticle–substrate interfaces by introducing an electron beam accelerated at a series of voltages between 18 and 20 keV and between 8.1 and 8.5 keV, respectively. For the latter case for example, as shown in Figure 1a, BSEs are generated predominantly within the nanoparticles at 8.1 keV, while more BSEs are generated from both the nanoparticles and the substrate at 8.5 keV (see Figure 1b). After the lower-acceleration SEM image is subtracted by the higher-acceleration SEM images, buried interfaces can be revealed.

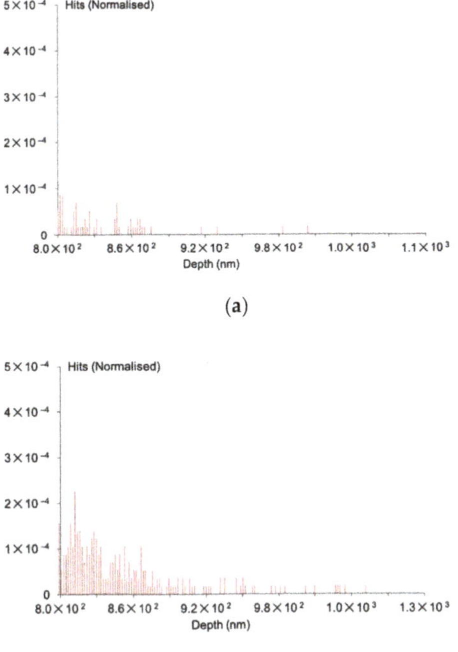

(a)

(b)

Figure 1. Backscattered electrons generated by the electron beam impacted on the nanoparticles at the acceleration voltages of (**a**) 8.1 and (**b**) 8.5 keV. These histograms are simulated by CASINO program [20].

5. Nanoparticles Synthesised by X-ray Radiolysis

The nanoparticles, #1 and 6, were imaged as shown in Figure 2a–h, respectively. These images were produced after subtracting two SEM images, which have been taken using different acceleration voltages of 18 and 20 keV. These images need to be aligned, which was carried out by adjusting the positions of the nanoparticles within the orange box shown in each image. The colour changes from magenta to green indicate that there are defects or vacancies within the subtracted image. The magenta colour shown at the edge of the particles in these images is due to its spherical shape as there is no intimate contact between the edge of the particles and the substrate. In addition, the bright and dark regions represent the number of BSEs generated to be more and less, respectively.

Figure 2a shows almost white and bright contrast at the nanoparticle–substrate interfaces, indicating that the interfaces are uniformly formed to generate a sufficient number of BSEs. Some arm-shaped regions appeared in magenta and green colours, indicating BSEs are generated above and below the interfaces, respectively. The magenta- and green-coloured features may indicate that the arm regions of the nanoparticles can be detached by voids. This may suggest that these arms can be formed once the main body of the nanoparticle (the middle region) is formed.

Figure 2b shows the nanoparticles synthesised on a n-doped Si(001) substrate with a sheet resistance of 1~10 $\Omega\cdot$cm. The size of the nanoparticles is found to be slightly randomised but maintains elongated shapes as seen in Figure 2a. The nanoparticle–substrate interfaces show broad distributions of contrast. This indicates that some nanoparticles with white bright interfaces are formed in the same manner with those synthesised on the Si/SiO$_2$. However, additional nanoparticles may have moved to form clusters, possibly due to the conductivity of the substrate.

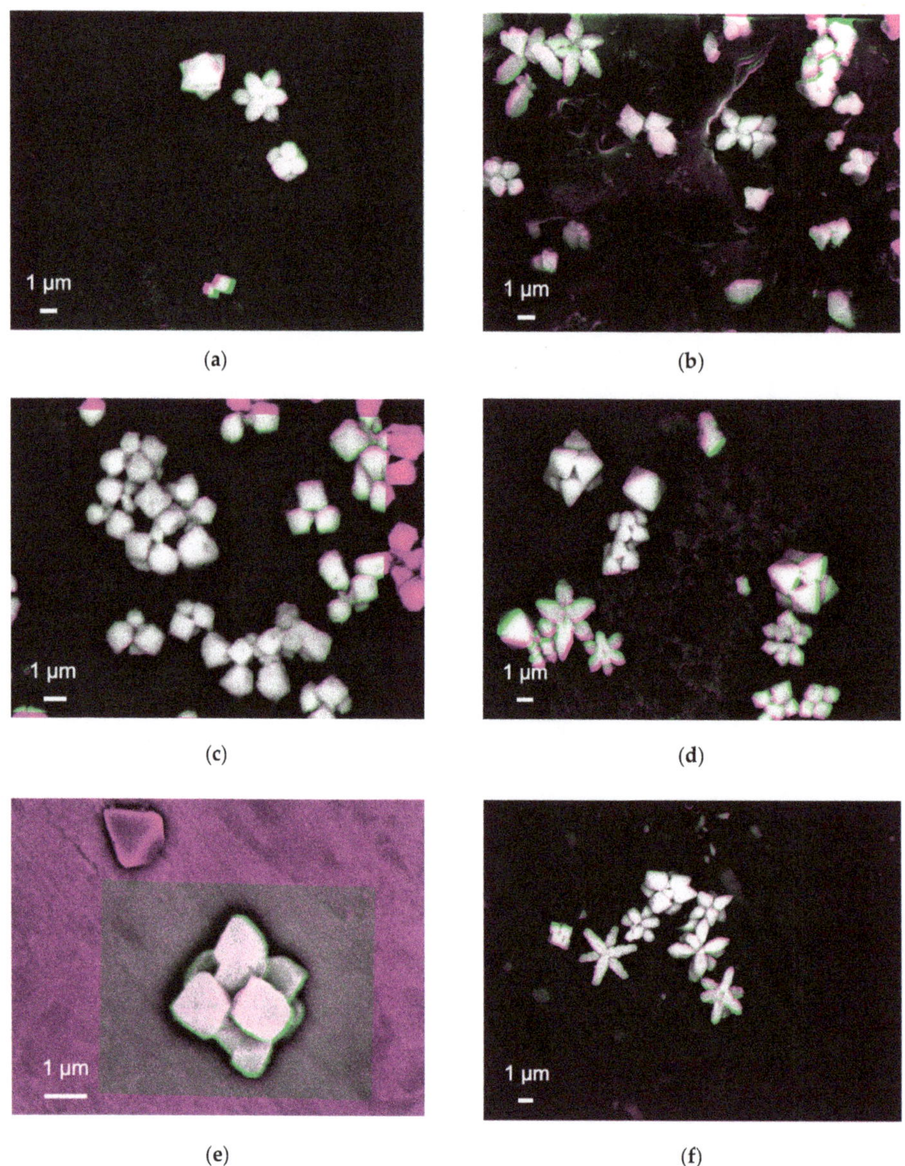

(a)

(b)

(c)

(d)

(e)

(f)

Figure 2. *Cont.*

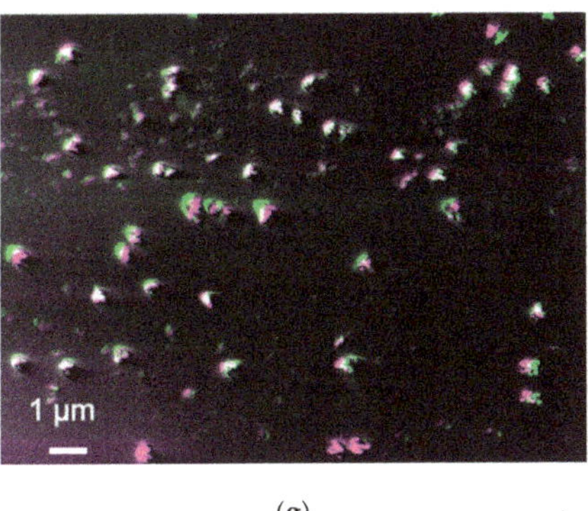 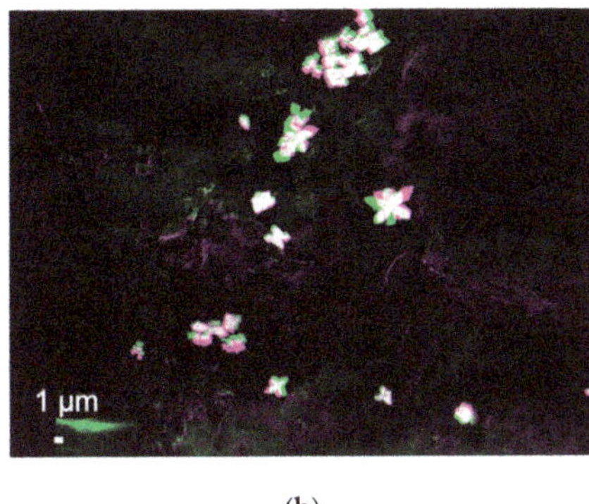

(g)   (h)

Figure 2. Processed images of the nanoparticle–substrate interfaces after subtracting two images taken at the acceleration voltages of 18 and 20 keV on the samples grown on (**a**) Si/SiO$_2$, (**b**) *n*-Si(001), (**c**) *p*-Si, (**d**) *n*-Si(111), (**e**) Ni, (**f**) Al, (**g**) 128° Y-cut LiNbO$_3$ and (**h**) PTFE substrates.

By replacing the substrate with *p*-doped Si(001) with a sheet resistance of 1~20 $\Omega\cdot$cm, the clustering of the nanoparticles is slightly suppressed by increasing the separation between the nanoparticles as shown in Figure 2c. The shape also becomes square like. The interfaces stay uniform. Their elongation is recovered by synthesising them on *n*-doped Si(111) as shown in Figure 2d, promoting triangular-shaped particles with closer clustering like those on *n*-Si(001).

On the other hand, the nanoparticles synthesised on the metallic Ni substrate show white bright contrast with magenta colour only without any arms as shown in Figure 2e. The shape and size of the nanoparticles are found to be almost cubic with three-fold symmetry as observed for the Si substrates as described above. Similar structures with more elongation are observed for the Al substrate as shown in Figure 2f. Randomly formed nanoparticles are observed for those synthesised on a 128° Y-cut LiNbO$_3$ substrate (see Figure 2g). In addition, some distorted particles are observed as immobilised on the LiNbO$_3$ substrate. They may be due to the chemical interactions with Cu(COOCH$_3$)$_2$ and difference in crystallinity between LiNbO$_3$ and cuprates. They become elongated on Al and polytetrafluoroethylene (PTFE) substrates. These results indicate that secondary electrons from the substrates by the X-ray introduction may contribute to the nucleation, growth, and aggregation of nanoparticles. It should be noted that all these samples maintain consistent interfaces.

Since prominent elongated arm-like features were obtained for those synthesised on Si/SiO$_2$ substrates, we further imaged nanoparticles synthesised under a series of pH between 7 and 9. Figure 3a shows almost white bright contrast at the nanoparticle–substrate interfaces for the Y-shaped nanoparticle, confirming that the interface is uniformly formed to generate a sufficient number of BSEs. There are some minor distributions in colour, where some voids exist at the interface. The inverse Y-shaped nanoparticle is found to be formed on the Y-shaped one as the inverse Y-shaped one has darker contrast, indicating the corresponding interface generates less BSEs, i.e., possible presence of voids.

A similar interfacial structure is observed by increasing pH to 9 as shown in Figure 3b. The contrast between the Y-shaped and inverse Y-shaped nanoparticles become weaker, meaning the interface for the latter one reduces the voids, i.e., the formation of a uniform interface. At the same time, each arm becomes more granular than that for the sample synthesised at pH = 8. This means the nanoparticle with the two overlapping Y-shapes is formed by clustering small circular particles to grow along six crystalline facets.

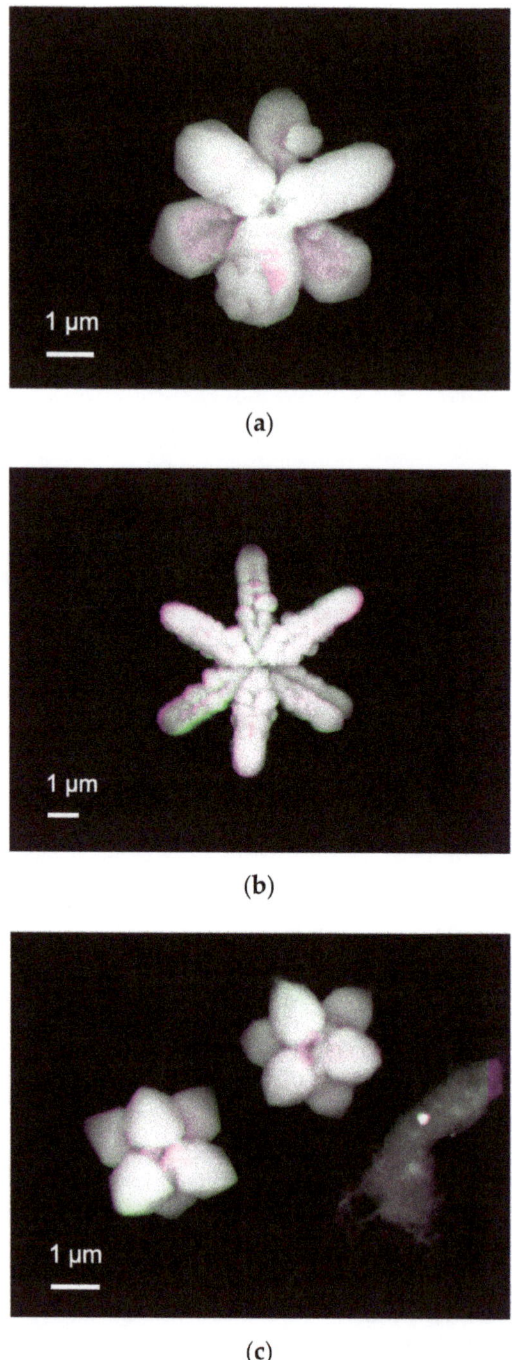

(a)

(b)

(c)

Figure 3. Processed images of the nanoparticle–substrate interfaces after subtracting two images 18 and 20 keV on the samples grown on Si/SiO$_2$ at pH = **(a)** 8, **(b)** 9, and **(c)** 7.

For pH = 7, the growth along the six facets becomes weaker to form less elongated arms as seen in Figure 3c. Again, the bottom Y-shaped nanoparticles have stronger adhesion to the substrate than the inverse Y-shaped one. By increasing pH to 11, the nanoparticles become almost like a sphere by attaching only the bottom centre of them. These results suggest that these cuprates are formed from triangular seed crystals, followed by preferred facet growth along the three directions. These cuprates grow in a Y-shape, whose arm length depends on the substrate and pH, which control the mobility of the seed crystals. By rotating 60° the Y-shaped cuprates to overlap each other, two of them can form a hexagonal structure [5]. These results shown here indicate that the composition and crystal shape of the synthesised and immobilised cupric nanoparticles are dependent on the conductivity of the substrates and pH of the liquid solutions. The formation of synthesised crystal can be modified when the relative order of the surface energies is altered or when the crystal growth along certain directions is selectively hindered. These results suggest that the selection of surface crystal structures and the electronic states of substrates play a dominant role in controlling the synthesis of nanoparticles.

6. Nanoparticles Synthesised by the Sol-Gel Method and Their Aqueous Suspensions

Suspensions prepared by the nanoparticles created by the sol-gel method were imaged as shown in Figure 4a–d, respectively, while some additional images were taken for a sample (i.e., Tug 3) as shown in Figure 5. These images were again produced after subtracting two SEM images, which were taken using different acceleration voltages of 8.1 and 8.5 keV. On the other hand, the results of the DLVO potential calculation are shown in Figure 6.

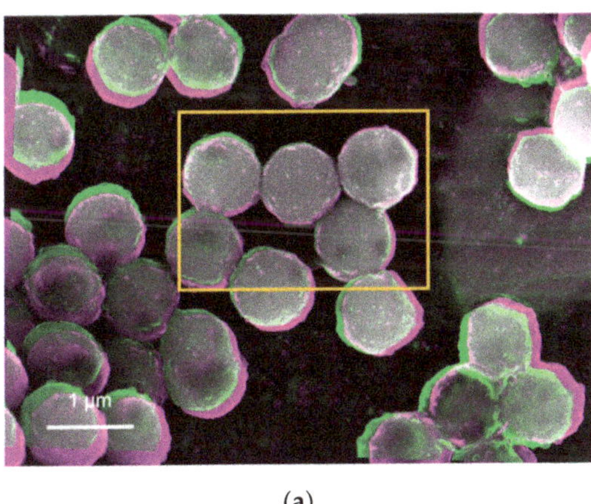

(a)

Figure 4. *Cont.*

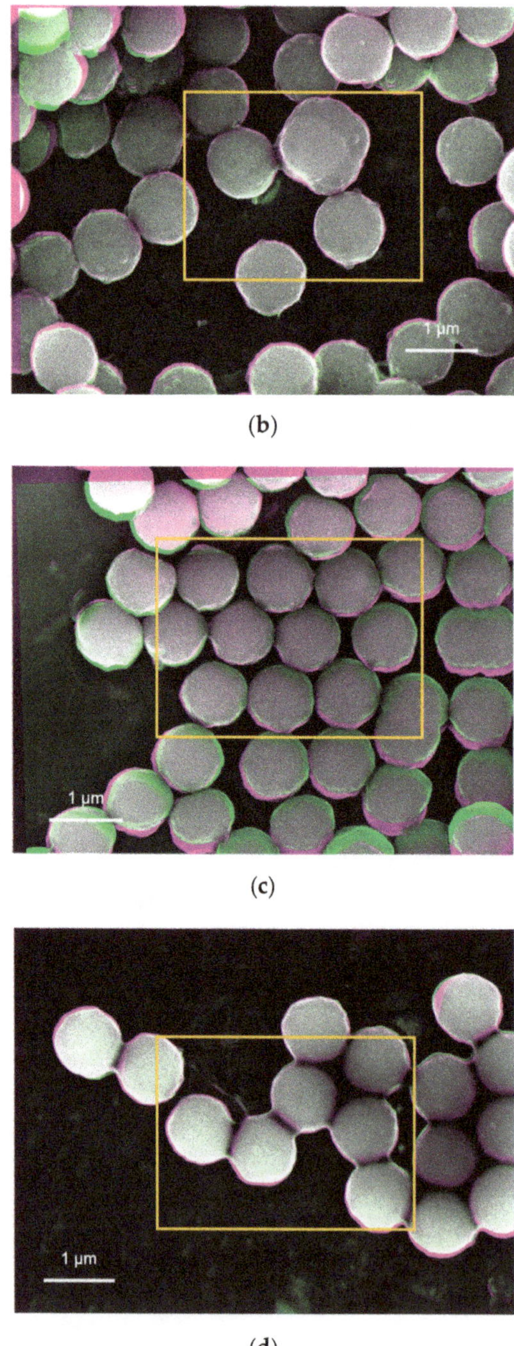

Figure 4. Processed images of the nanoparticle–substrate interfaces after subtracting two images taken at the acceleration voltages of 8.5 and 8.1 keV on the samples of Tug (**a**) 2, (**b**) 3, (**c**) 5, and (**d**) 6.

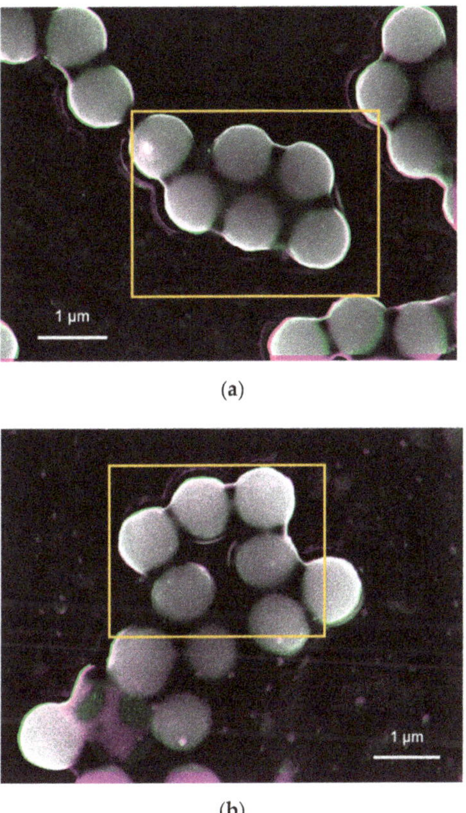

(**a**)

(**b**)

Figure 5. Processed images of the nanoparticle–substrate interfaces after subtracting two images taken at the acceleration voltages of 8.5 and 8.1 keV on the Tug 3 sample with different areas (**a**,**b**).

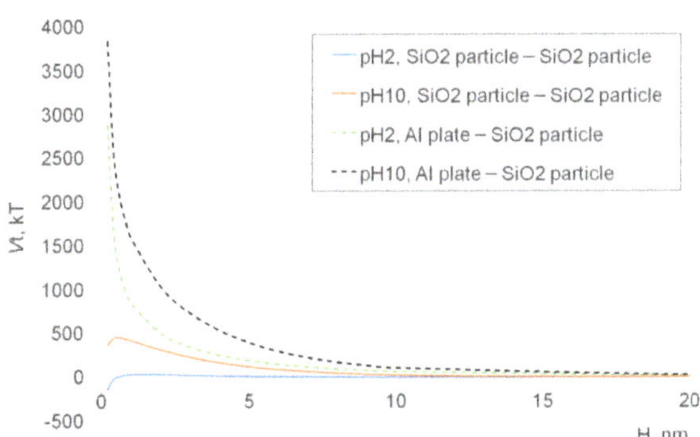

Figure 6. DLVO total potential energies between silica–silica particles with the radius of 280 nm as well as between the Al stub plate and silica particles at pH 2 or 10 and 1×10^{-2} M KNO_3 aqueous solution at 25 °C, as a function of interparticle separation distance. The unit of the potential, Vt is the thermal energy, $k_B T$ [J].

In the processed images shown in Figures 4 and 5, in general, the edges of these particles show bright/white colour, indicating that they generate more BSEs as compared with the other interfaces. This suggests that the particles are pinned by these edges due to the capillary force that is often expressed by the neck shape structure between a particle and a plate in the presence of a small amount of water [22–25]. On the other hand, some edges are shown in green or magenta colour, indicating that they are detached from the substrate due to repulsive force that can be explained by the electrostatic interactions between particles and/or particle and the substrate/stub.

In Figure 4a, some nanoparticles contain grey spots (less BSEs), indicating there are some defects or vacancies formed at the nanoparticle, substrate, or their interface. Minor bright edges are observed while other edges in green or magenta are also seen in Tug 2 (Figure 4a), indicating that the nanoparticles are deposited on the substrate with some instability due to electrostatic repulsion among the highly charged silica particles at pH 10 (Tug 2), as the repulsive DLVO potential interaction among them is shown in Figure 6 (pH 10 SiO_2 particle–SiO_2 particle; pH 10 Al plate–SiO_2 particle). It is more noticeable with Tug 2 prepared at pH 10 (Figure 4a) than Tug 5 at pH 2 (Figure 4c), and this agrees with the DLVO potential calculation shown in Figure 6 (pH 10 SiO_2 particle–SiO_2 particle vs. pH 2 SiO_2 particle–SiO_2 particle) explaining the repulsive interaction at pH 10 while the attractive interaction at pH 2 between SiO_2 particles. Here, it is worth mentioning that the capillary force can be stronger than the attractive DLVO forces [26] (mainly van der Waals force in our case) in order to keep those particles on the substrate/stub while the repulsive DLVO forces (electrostatic force in our case) can influence the stability of particle deposition on the substrate/stub. Similarly, Figure 4c shows almost uniform interfaces at the middle of the nanoparticles but with some edge defects as shown in the green colour. Tug 5 (Figure 4c, pH 2) has a more flat structure in comparison with Tug 2 (Figure 4a, pH 10) that forms multilayer disposition at the same particle concentration of 0.1 vol.%, indicating more stable adhesion between the silica particles and substrate, as the attractive DLVO potential interaction among them at pH 2 shows in Figure 6 (pH 2 SiO_2 particle-SiO_2 particle).

Figure 4b shows similar interfacial contrast but with darker regions at the edges of the nanoparticles whose suspension was prepared at pH 10 and 0.01 vol.% silica. Furthermore, some interparticle spots in green colour are observed. This indicates that the silica nanoparticles do not have a perfect spherical shape or have repulsive interactions to separate the nanoparticles from each other. On the other hand, Figure 4d shows no interparticle spots in the green colour, indicating attractive interactions between particles as agreed with the DLVO potential calculation (Figure 6, pH 2 SiO_2 particle–SiO_2 particle). The more uniform colour regions are observed in the Tug 3 (Figure 4b) and Tug 6 (Figure 4d) prepared at 0.01 vol.% silica, where there is a higher amount of moisture content than 0.1 vol.% (Tug 2, Figure 4a and Tug 5, Figure 4c), and it indicated that silica nanoparticles are firmly deposited on the stub by the capillary force with the residual water.

In terms of the interactions between the aluminium stub and silica particles, they are more repulsive at pH 10 and 0.1 vol.% silica particles (Tug 2) than at pH 2 and 0.1 vol.% silica particles (Tug 5), as shown in Figure 4a,c and agreed with the highly positive DLVO potential energies in the former sample condition (Figure 6, pH 10 vs. pH 2) and literature [16]. It can be also explained by the isoelectric point (IEP) of silica, which is around pH 2, where silica particles can coagulate while pH 10 is where the solution pH is far away from IEP and thus silica particles repel each other [16]. Comparing between 0.1 vol.% (Tug 2, pH 10) and 0.01 vol.% (Tug 3, pH 10), the former solid concentration provides more repulsive–instable interactions. It can be explained by the presence of a higher number of particles that are deposited on the aluminium stub with the repulsive nature of interactions.

SEM images were also taken in the other three different areas for each sample to confirm the representativity of the observed images shown in Figure 5 that are the processed images of Tug 3 (pH 10, 0.01 vol.%) as an example. No apparent difference is observed for all the four samples, apart from a minor difference in the particle orientation for the Tug 3. It can be explained by a non-even particle distribution as shown in Figure 5.

7. Summary

By using our non-destructive imaging method, we imaged nanoparticles synthesised by X-ray radiolysis and the sol-gel method. The X-ray radiolysis is found to initiate the formation of a triangular seed crystal, followed by growth along three facet directions. The sol-gel method, on the other hand, forms spherical nanoparticles, which are pinned to the substrate at the interface and clustered randomly. These crystallisation, deposition, and aggregation processes can be controlled by the substrates, pH, and density as expected and agreed with colloidal DLVO theory. Our imaging method can offer an ideal feedback to achieve precise control of the synthesis processes.

Author Contributions: All authors contributed to write this article. A.Y. synthesised nanoparticles and A.O. prepared nanoparticle suspensions. K.E. made all the imaging. A.H. developed the imaging method and analysed the images with K.E., A.O. and N.L.H. performed DLVO potential calculation. All authors have read and agreed to the published version of the manuscript.

Funding: The imaging work is partially supported by Japan Society for the Promotion of Science (JSPS)-Engineering and Physical Sciences Research Council (EPSRC) Core-to-core programme (EP/M02458X/1) and Japan Science and Technology Agency (JST) Core Research for Evolutional Science and Technology (CREST) (JPMJCR17J5). The authors thank financial and technical support by JEOL UK to develop the non-destructive imaging. The nanoparticle synthesis and characterisation were partially supported by European Soft Matter Infrastructure (EUSMI) and the visiting professorship of the Institute for Solid State Physics in the University of Tokyo, respectively.

Institutional Review Board Statement: Not applicable.

Informed Consent Statement: Not applicable.

Data Availability Statement: Data is contained within the article and available on request with following the guideline set by the University of York (UK).

Acknowledgments: A.O. wishes to acknowledge his thanks to Schofield for the silica particle synthesis and Hamane for TEM imaging used for size measurement of silica nanoparticles.

Conflicts of Interest: The authors declare no conflict of interest.

References

1. Hirohata, A.; Frost, W.; Hillebrands, B. Structural and electro-magnetic characterisation. In *Magnetic Materials: Fabrication, Characterization and Application*; Elsevier: Amsterdam, The Netherlands, in press.
2. Hirohata, A.; Yamamoto, Y.; Murphy, B.A.; Vick, A.J. Non-destructive imaging of buried electronic interfaces using a decelerated scanning electron beam. *Nat. Commun.* **2016**, *7*, 12701. [CrossRef] [PubMed]
3. Jackson, E.; Wu, Y.; Frost, W.; Kim, J.-Y.; Samiepour, M.; Elphick, K.; Sun, M.; Kubota, T.; Takanashi, K.; Ichinose, T.; et al. Non-destructive imaging for quality assurance of magnetoresistive random-access memory junctions. *J. Phys. D Appl. Phys.* **2020**, *53*, 014004. [CrossRef]
4. Jackson, E.; Sun, M.; Kubota, T.; Takanashi, K.; Hirohata, A. Chemical and structural analysis on magnetic tunnel junctions using a decelerated scanning electron beam. *Sci. Rep.* **2018**, *8*, 7585. [CrossRef] [PubMed]
5. Yamaguchi, A.; Okada, I.; Sakurai, I.; Izumi, H.; Ishihara, M.; Fukuoka, T.; Suzuki, S.; Elphick, K.; Jackson, E.; Hirohata, A.; et al. Controllability of cupric particle synthesis by linear alcohol chain number as additive and pH control in cupric acetate solution using X-ray radiolysis. *J. Synchrotron Radiat.* **2019**, *26*, 1986. [CrossRef] [PubMed]
6. Van Blaaderen, A.; Vrij, A. Synthesis and Characterization of Colloidal Dispersions of Fluorescent, Monodisperse Silica Spheres. *Langmuir* **1992**, *8*, 2921–2931. [CrossRef]
7. Derjaguin, B.V.; Landau, L. Theory of the Stability of Strongly Charged Lyophobic Sols and of the Adhesion of Strongly Charged Particles in Solutions of Electrolytes. *Acta Physicochim.* **1941**, *14*, 633–662. [CrossRef]
8. Verwey, E.J.W.; Overbeek, J.T.C. *Theory of the Stability of Lyophobic Colloids*; Elsevier Publishing Company Inc.: New York, NY, USA, 1948.
9. Gotoh, K.; Inoue, T.; Tagawa, M. Adhesion of nylon particles to a quartz plate in an aqueous solution and their removal by electro-osmosis. *Colloid Polym. Sci.* **1984**, *262*, 982–989. [CrossRef]
10. Kallay, N.; Barouch, E.; Matijevic, E. Diffusional detachment of colloidal particles from solid/solution interfaces. *Adv. Colloid Interface Sci.* **1987**, *27*, 1–42.
11. Furusawa, K.; Anzai, C. Heterocoagulation behaviour of polymer latices with spherical silica. *Colloids Surf.* **1992**, *63*, 103–111. [CrossRef]

12. Otsuki, A.; Dodbiba, G.; Fujita, T. Two-Liquid Flotation for Separating Mixtures of Ultra-Fine Rare Earth Fluorescent Powders for Material Recycling—A Review. *Colloids Interfaces* **2018**, *2*, 7. [CrossRef]

13. Otsuki, A.; Hayagan, N.L. Zeta potential of inorganic fine particle—Na-bentonite binder mixture systems. *Electrophoresis* **2020**, *41*, 1405–1412. [CrossRef] [PubMed]

14. Otsuki, A. Coupling colloidal forces with yield stress of charged inorganic particle suspension: A review. *Electrophoresis* **2018**, *39*, 690–701. [CrossRef] [PubMed]

15. Otsuki, A.; Dodbiba, G.; Fujita, T. Two-liquid flotation: Heterocoagulation of fine particles in polar organic solvent. *Mater. Trans.* **2007**, *48*, 1095–1104. [CrossRef]

16. Otsuki, A.; Barry, S.; Fornasiero, D. Rheological studies of nickel oxide and quartz/hematite mixture systems. *Adv. Powder Technol.* **2011**, *22*, 471–475. [CrossRef]

17. Cherepy, N.J.; Shen, T.H.; Esposito, A.P.; Tillotson, T.M. Characterization of an effective cleaning procedure for aluminum alloys: Surface enhanced Raman spectroscopy and zeta potential analysis. *J. Colloid Interface Sci.* **2005**, *282*, 80–86. [CrossRef]

18. Elimelech, M.; Gregory, J.; Jia, X. *Particle Deposition & Aggregation—Measurement, Modelling and Simulation*; Butterworth-Heinemann: Woburn, MA, USA, 1995; pp. 33–67.

19. Visser, J. On Hamaker constants: A comparison between Hamaker constants and Lifshitz-van der Waals constants. *Adv. Colloid Interface Sci.* **1972**, *3*, 331–363. [CrossRef]

20. Bergstrom, L. Hamaker constants of inorganic materials. *Adv. Colloid Interface Sci.* **1997**, *70*, 125–169.

21. Drouin, D.; Couture, A.R.; Joly, D.; Tastet, X.; Aimez, V.; Gauvin, R. CASINO V2.42—A Fast and Easy-to-use Modeling Tool for Scanning Electron Microscopy and Microanalysis Users. *Scanning* **2007**, *29*, 92. [CrossRef]

22. Flury, M.; Aramrak, S. Role of air-water interfaces in colloid transport in porous media: A review. *Water Resour. Res.* **2017**, *53*, 5247–5275. [CrossRef]

23. Xu, Q.; Jensen, K.E.; Boltyanskiy, R.; Sarfati, R.; Style, R.W.; Dufresne, E. Direct Measurement of Strain-dependent Solid Surface Stress. *Nat. Commun.* **2017**, *8*, 555. [CrossRef]

24. Quang, T.S.B.; Leong, F.Y.; An, H.; Tan, B.H.; Ohl, C.D. Growth and wetting of water droplet condensed between micron-sized particles and substrate. *Sci. Rep.* **2016**, *6*, 30989. [CrossRef] [PubMed]

25. Leite, F.L.; Bueno, C.C.; Da Roz, A.L.; Ziemath, E.C.; Oliveira, O.N., Jr. Theoretical Models for Surface Forces and Adhesion and Their Measurement Using Atomic Force Microscopy. *Int. J. Mol. Sci.* **2012**, *13*, 12773–12856. [CrossRef] [PubMed]

26. Sharma, P.; Flury, M.; Zhou, J. Detachment of colloids from a solid surface by a moving air-water interface. *J. Colloid Interface Sci.* **2008**, *326*, 143–150. [CrossRef] [PubMed]

 *materials*

Article

Thermochemical Route for Extraction and Recycling of Critical, Strategic and High Value Elements from By-Products and End-of-Life Materials, Part I: Treatment of a Copper By-Product in Air Atmosphere

Ndue Kanari [1,*], Eric Allain [1], Seit Shallari [2], Frederic Diot [1], Sebastien Diliberto [3], Fabrice Patisson [3] and Jacques Yvon [1]

[1] GeoRessources Laboratory, UMR 7359 CNRS, CREGU, Université de Lorraine, 2, rue du doyen Roubault, BP 10162, 54505 Vandoeuvre-lès-Nancy, France; ericgallain@gmail.com (E.A.); frederic.diot@univ-lorraine.fr (F.D.); jacques.yvon@univ-lorraine.fr (J.Y.)
[2] Agricultural University of Tirana, Faculty of Agriculture and Environment, 1029 Tirana, Albania; seitshallari@gmail.com
[3] Institut Jean Lamour, UMR 7198 CNRS, Labex DAMAS, Université de Lorraine, Campus Artem, 2 allée André Guinier, BP 50840, 54011 Nancy, France; sebastien.diliberto@univ-lorraine.fr (S.D.); fabrice.patisson@univ-lorraine.fr (F.P.)
* Correspondence: ndue.kanari@univ-lorraine.fr; Tel.: +33-372-744-530

Received: 26 April 2019; Accepted: 15 May 2019; Published: 17 May 2019

Abstract: Development of our modern society requests a number of critical and strategic elements (platinum group metals, In, Ga, Ge …) and high value added elements (Au, Ag, Se, Te, Ni …) which are often concentrated in by-products during the extraction of base metals (Cu, Pb, Zn …). Further, recycling of end-of-life materials employed in high technology, renewable energy and transport by conventional extractive processes also leads to the concentration of such chemical elements and their compounds in metallurgical by-products and/or co-products. One of these materials, copper anode slime (CAS), derived from a copper electrolytic refining factory, was used for this study. The sample was subjected to isothermal treatment from 225 to 770 °C under air atmosphere and the reaction products were systematically analyzed by scanning electron microscopy through energy dispersive spectroscopy (SEM-EDS) and X-ray diffraction (XRD) to investigate the thermal behavior of the treated sample. The main components of the anode slime ($CuAgSe$, $Cu_{2-x}Se_yS_{1-y}$, Ag_3AuSe_2) react with oxygen, producing mostly copper and selenium oxides as well as Ag-Au alloys as final products at temperatures higher than 500 °C. Selenium dioxide (SeO_2) is volatilized and recovered in pure state by cooling the gaseous phase, whilst copper(II) oxide, silver, gold and tellurium remain in the treatment residue.

Keywords: critical and strategic materials; valuable metals; by-products; copper anode slime; thermal treatment; selenium dioxide

---

1. Introduction

Nowadays, besides the most common metals (Fe, Al, Cu, Zn, Ni …), a wide range of chemical elements (metals, non-metals, and metalloids) are in increasing demand and crucial for the development of renewable energies, the manufacture of electrical and electronic equipment, transportation, advanced industries, which constitute the foundation of innovative high technologies. Often suffering from a slight subjective and arbitrary classification, these elements known as rare elements, or sometimes as small, green, or high-tech metals, are commonly considered to be critical and strategic materials. According to criticality assessments, the number of critical materials (CRMs) for the European Union has grown from 14 CRMs in 2011 to 20 in 2014 and was augmented thereafter to reach 27 CRMs in 2017 [1].

While the importance of CRMs in our society and their inventory is clear, the fragility of their supply represents a real challenge for the French, European and global economies. In general, material savings, substitution and recycling (including eco-design) can mitigate the risk of anticipated shortage of critical metals. Critical and strategic materials are often used in unique applications with limited alternatives and substitutions [2]. Note that the actual and future demand for several of these materials are expected to strongly grow, especially for the production of electric and electronic devices, while the life-spans of many products such as computers and cell phone are decreasing rapidly [3]. This phenomenon, among others, leads to the generation of a huge waste stream (waste electrical and electronic equipment-WEEE), often designated as "e-waste" [4–7]. Its elevated contents in high added values elements (precious, strategic, critical, rare and rare earths elements) make this waste stream a very attractive "urban mine" for the extraction of these components.

But paradoxically, the rate of recycling of CRMs from end-of-life materials is low compared to that of base metals. The presence of several metals of different chemical nature in increasingly miniaturized products, and incorporated in complex matrixes encompassing ceramics, resins, plastics, and especially halogenated substances (based on chlorine and bromine) makes their recycling process quite complex. A spectrum of contributions focused on current state-of-the-art research on metal recycling from wastes was introduced in a special issue (Valuable Metal Recycling) of *Metals* [8]. An extensive overview for the recovery of high value metals from electronic waste and spent catalysts using improved pyrometallurgical and hydrometallurgical technologies was given by Ding et al. [9]. It was mentioned that smelting is beneficial to segregate and concentrate the precious metals from e-waste in the by-products. In a recent investigation, Avarmaa et al. [10] demonstrated that during smelting, copper acts as collector metal for gold, silver, palladium and platinum contained in the end-of-life electronic devices. One may emphasize that most of the targeted elements (Au, Ag, Se, Te, platinum group metals, etc.) are recovered in anode slime from primary copper extraction from its sulfide ores and concentrates. In summary, recycling of 'e-waste' by pyrometallurgical route in the existing copper metallurgy leads to the concentration of these elements in by-and/or co-products, mostly in the anode slime of copper electrorefining.

A number of research reports were devoted to extraction and recovery of specific elements from copper anode slime (CAS). The most typical investigations concerned CAS treatment by H_2SO_4-O_2 to recover Cu followed by thiourea leaching for Ag extraction [11] or leaching in a sulfuric and nitric acid medium which was used for silver nanoparticles synthesis [12]. A nitric-sulfuric acid mixture was also tested by Li et al. [13] for the recovery of Cu, Ag and Se from a CAS sample, while Xiao et al. [14] reported an oxidative pretreatment by thiosulfate leaching followed by an electrodeposition process for the selective recovery of Ag. Dissolution of selenium in NaOH, after sulfuric acid decopperization of CAS, was also reported [15]. However, little recent information is available for the thermal behavior of CAS under an oxidizing atmosphere. In this frame, the work presented in this article deals with the treatment in an air atmosphere of a CAS sample generated by a European copper plant. Particular attention was payed to the speciation and physical characterization of the obtained products (residues and condensates) at various temperatures to determine the reaction mechanisms and to identify process steps for a possible selective separation of the CAS components.

This research work is an extension of our investigations on the concentration and extraction of a broad range of valuable metals from raw materials and solid industrial residues by thermal route [16–20] as well as on the recycling and valorization of co-products and wastes via chemical synthesis of new materials [21–24].

At least two of the researched elements of this investigation, selenium and tellurium, belong also to the scattered elements category; i.e., which exist at low content, in a decentralized state and rarely form independent minerals in nature [25]. Both elements are used in thin films (CIGS—copper indium gallium selenide and CdTe—cadmium telluride) and used in second-generation modules of photovoltaic panels [26–28]. Efficient extraction and recycling of these elements from by-products, end-of-life solar photovoltaics and other wasted materials will be a main research focus in the future to meet the volume of industrial demand.

2. Materials and Methods

A sample of copper anode slime provided from a refining factory available as a powdered solid with a mean particles size less than 30 μm was used for this study. The sample composition was determined by diverse analytical methods, but only the results of scanning electron microscopy-energy dispersive spectroscopy (SEM-EDS) and X-ray diffraction (XRD) analyses were selected to be presented in this paper (see Section 3). SEM examinations were performed by using a HITACHI S-4800 device (Hitachi Ltd., Tokyo, Japan) equipped with an EDS elemental analysis microprobe. SEM imaging and chemical microanalysis were performed mostly at an accelerating voltage of 15 kV. The samples were carbon coated prior to analysis. XRD was performed on samples using a Bruker D8 Advance device (Bruker, Karlsruhe, Germany), (Co Kα1 radiation, λ = 1.789 Å), under 35 kV and 45 mA operating conditions. The acquisition of diffraction patterns was performed between 3° and 64° (2θ) range at step scan 0.034° with 3 s per step. The XRD-patterns were analyzed thanks to the DIFFRAC.EVA software (version 5, Bruker) and PDF-2 release 2011 database.

Isothermal experimental tests for the CAS sample treatment were conducted in horizontal set-ups, including a system of static tubular furnaces that were able to reach 1600 °C. To achieve this, a pre-weighted sample of several grams was introduced directly into the furnace preheated at the desired temperature. When the dwell time was reached, the sample was removed from the furnace and cooled down to room temperature. Air was used as a flowing gas and it also performed the oxidation of CAS components. The outlet gases were cooled at room temperature, leading to the condensation of the vapor phase and the recovery of a solid condensate. Initial CAS sample and solid products obtained from CAS thermal treatment were examined by visible microscopy, SEM-EDS and XRD. The advantage of using SEM-EDS analysis here is the ability to gather punctual information about elemental content, phases differentiation, and about the morphological and textural evolution of the thermally treated samples, all this contributes to a better understanding of the involved reaction mechanism and processing steps.

3. Results

3.1. Elemental and Mineralogical Analysis of CAS Sample

Several research reports [11–15,29–32] found in the literature revealed that the copper anode slime is a quite complex material in elemental and mineralogical composition, as well as in morphological aspects. This complexity resulted from the nature of copper raw materials, operating conditions of copper extractive processes (smelting, converting, fire- and electro-refining) and from the decopperization step applied to treat CAS. By the same token, a general SEM image of the used CAS sample for this study given in Figure 1a shows a particular morphology of this solid. Besides the spherical particles, there are many particles that are irregularly and shell shaped. As clearly shown by the SEM image, most particles are corroded, resulting in three-dimensional lens-like forms and other ones with plenty of cavities and pores.

The general EDS spectrum of the CAS (Figure 1b) indicated a multi-elemental composition with a descending peak intensity order as follows: Cu, Se, Ag, O, S, Au, Te, Si, Cl. There is probably also a carbonaceous matter in the CAS sample, but it was difficult to distinguish from the carbon coating used to make the sample conductive before SEM-EDS analysis.

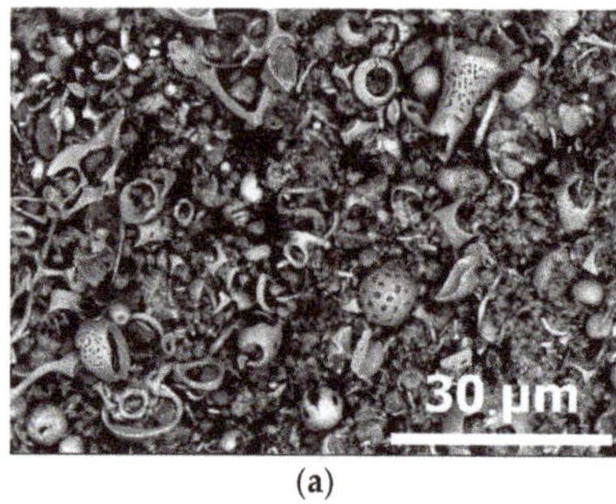
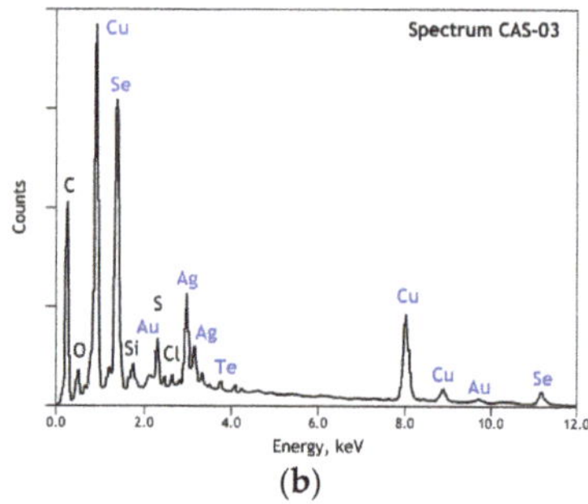

(a)   (b)

Figure 1. SEM-EDS results of initial copper anode slime: (**a**) General view (backscattered electron micrograph) of the used sample; (**b**) Overall EDS analysis of the used sample.

The morphology of the CAS sample showed in Figure 2 and SEM-EDS microanalysis data summarized in Table 1, indicates that there are particles containing mostly Cu-Se (spot n° 1 and 2), Ag-Se-Cu-Te (spot n° 3) and Ag-Se-Cu-Au-Te (spot n° 4); gold is found mostly in the finest particles. There is also some sulfur in almost all particles, while the small amount of oxygen is bound to silicon. This peculiar state of CAS particles together with their composition can be explained by the preceding impact of at least two factors: (i) the behavior the Cu, Ag, Au, Se, Te, S, O and others elements compounds in fire refining process of blister copper and (ii) the chemical and electrochemical reactivity of these compounds towards electrolyte during electrorefining of anode copper. The fire refining process, with the goal of removing residual oxygen and sulfur as well as other impurities from blister copper before casting into anode, is achieved at 1200–1250 °C.

As could be expected, the residual oxygen and sulfur can be associated with copper forms as Cu_2O and Cu_2S, having higher melting points than copper. One may deduce that during the cooling of casted Cu-anodes, the Cu_2O and Cu_2S particles, which are first to be solidified, can serve as heterogeneous nuclei for further solidification around these nucleation sites. The phase diagrams for copper with chalcogen (e.g., S, Se, or Te) show that there are miscibility gaps in the liquid phase [33] that will allow the solidification of droplets with the highest melting points followed by further solidification of low melting point components, thereby resulting in a layered structure and in a significant difference between the core and the periphery of the obtained solid. During copper electrorefining, the copper oxide (CuO) contained in mixed (Cu, Se, Te, S, O …) particles as well as common impurities (most probably oxides of As, Sb and Bi) are dissolved, besides electrochemical dissolution of Cu°, resulting in the final configuration of CAS. These observations are in good agreement with other research works reported in the literature [29–32].

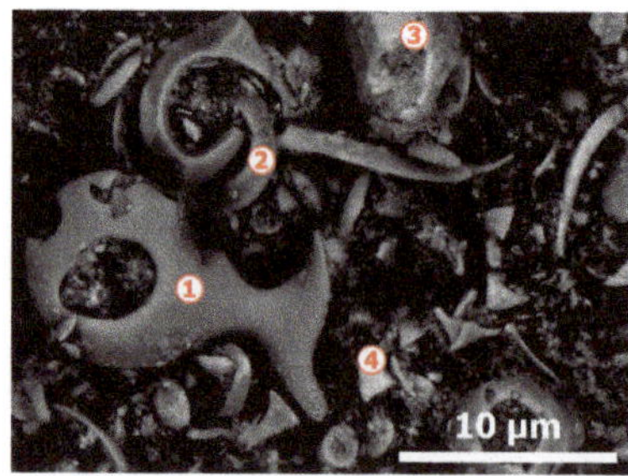

Figure 2. Detailed aspects (backscattered electron micrograph) of initial CAS as revealed by SEM-EDS.

Table 1. Elemental composition of initial CAS analyzed by SEM-EDS.

Elements	Spot n° 1		Spot n° 2		Spot n° 3		Spot n° 4	
	wt % [1]	at % [1]	wt %	at %	wt %	at %	wt %	at %
O	-	-	-	-	-	-	0.49	2.55
Si	-	-	-	-	-	-	1.01	3.00
S	2.01	4.12	2.87	6.01	1.72	4.36	1.46	3.79
Cl	-	-	-	-	0.63	1.45	-	-
Cu	72.20	74.48	58.55	61.88	22.21	28.47	30.79	40.28
Se	25.78	21.40	35.48	30.17	32.57	33.59	20.34	21.42
Ag	-	-	3.11	1.93	40.88	30.89	27.05	20.84
Te	-	-	-	-	1.98	1.27	0.67	0.43
Au	-	-	-	-	-	-	18.19	7.68

[1] wt % and at % represent mass and atomic percentage, respectively.

XRD results of the used CAS are shown in Figure 3. The main crystallized phase is Eucairite (CuAgSe), confirming the microanalysis obtained by SEM-EDS analysis. However, this phase probably forms during the electro-refining process. According to Chen and Dutrizac [29,32], silver is in a metastable state in the copper anode and this can promote the chemical and/or electrochemical dissolution of Ag° during electrorefining. The resulting ion Ag$^+$ can react following Equation (1) by synthetizing the Eucairite phase. In addition, the authors stated that the particles kept the original morphologies of Cu$_2$Se [Cu$_2$(Se,Te)], although their composition has been changed. The conditions of the electrorefining process (mean values: T $\approx$ 60–65 °C; $\approx$ 180–185 g/L H$_2$SO$_4$; residence time of several days) will certainly favor such a reaction.

$$Ag^+ + Cu_2(Se,Te) \rightarrow AgCu(Se,Te) + Cu^+ \tag{1}$$

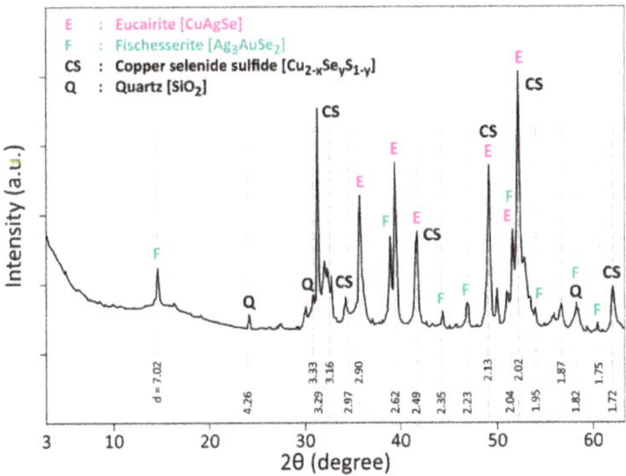

Figure 3. XRD patterns of CAS raw sample.

In the XRD patterns of CAS, there are some peaks corresponding to Cu$_2$Se, Cu$_2$S, CuSeS, as well as to non-stoichiometric ones with an absence of certain peaks due probably to crystallite orientation; for simplification, all of these phases are designated as copper selenide sulfide (Cu$_{2-x}$Se$_y$S$_{1-y}$). Regarding the Au-bearing phase, the XRD peaks of the obtained diffractogram match most frequently with Fischesserite (Ag$_3$AuSe$_2$). This phase composition (with substitution of Ag by Cu and Se by Te) is also confirmed by SEM-EDS analysis (spot n° 3 and spot n° 4 in Table 1). Finally, Quartz (SiO$_2$) completes the set of crystallized phases identified in the CAS sample.

3.2. Thermal Treatment of CAS in Air for 1 h

Several grams of the copper anode sample were used for isothermal tests performed between 225 °C and 770 °C in air atmosphere with a flow rate of 25 L/h. The evolution of the % mass loss (%ML) as a function of temperature is plotted in Figure 4. The %ML curve shape suggests that many reactions steps may occur in the explored temperature interval. The sample started to react with air oxygen (O_2) at temperatures higher than 225 °C, resulting in a mass gain of up to about 350 °C. From here, there is a continued mass loss up to around 700 °C, where the mass loss is stabilized.

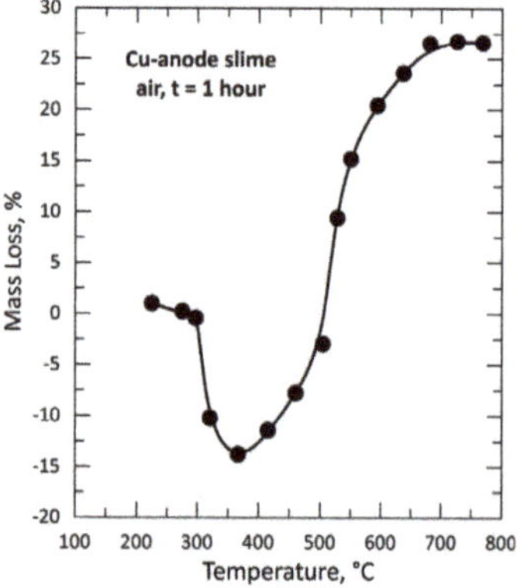

Figure 4. Evolution of the mass loss of the sample versus temperature during treatment of CAS in air for 1 h.

According to the composition of raw CAS sample, the main constituents which may strongly affect the %ML are Cu, Se and Ag combined as selenides of copper and silver. During their reactions with air (oxide formation) and knowing the atomic mass of these elements (Cu: 63.55; Se: 78.97 and Ag: 107.87, O: 16.00), it appears that no volatilization of these elements and/or their oxides occurred, at least up to 370 °C, since only a mass gain was recorded. The most volatile compound of the Cu-Se-Ag-O system is SeO_2, having a vapor pressure of 73 kPa at 320 °C [34] which is enough for its volatilization during the CAS treatment at this temperature. This is an indirect confirmation that no free SeO_2 is formed at T < 320 °C when CAS sample is heated in air. More consistent information about the interaction steps of CAS components with air is given in the subsequent section.

3.3. Analysis of the Reaction Products

The treatment residues issued from isothermal tests of CAS under air atmosphere were systematically examined by XRD technique and results are presented in Figures 5 and 6 for the treatment temperatures of 225–415 °C and 505–770 °C, respectively.

As shown in Figure 5, the X-ray diffractogram of the residue obtained at 225 °C is similar to that of the CAS raw sample. The XRD patterns of the sample heated 320 °C exhibits a fair broad diffraction band near to 6° (2θ), indicating a slight XRD disordered structure, at least, that of a reaction product which will be a precursor of further crystallized phases. Eucairite (CuAgSe) seems to be still stable at these temperatures, while the characteristic peaks of $Cu_{2-x}Se_yS_{1-y}$ and Ag_3AuSe_2 phases disappeared. Apparition of new crystallized phases such as: $[Cu_4O(SeO_3)_3]$, $[Cu_2O(SeO_3)]$ and Ag° is obvious from the XRD examination. One may hypothesize that $[Cu_4O(SeO_3)_3]$, which can be written also as $[4CuO{\cdot}3SeO_2]$, is the first crystallized phase issued from the reaction of copper selenide bearing substances with O_2, though other phases not detected by XRD may be the intermediate product of selenide oxidation by oxygen. Metallic silver is also identified in this treatment residue; note that the main XRD peaks of Ag° and Au° overlapped. Further increase in the treatment temperature (from 320 °C to 415 °C) of CAS sample leads to a decrease in the reflection intensities of CuAgSe and $Cu_4O(SeO_3)_3$ in the generated residue, while the relative intensity of $[Cu_2O(SeO_3)]$ peaks seems increased. As noted by Portnichenko et al. [35], copper(II)-oxoselenite $[Cu_2OSeO_3]$ has a cubic structure with a cell parameter a = 8.925 Å.

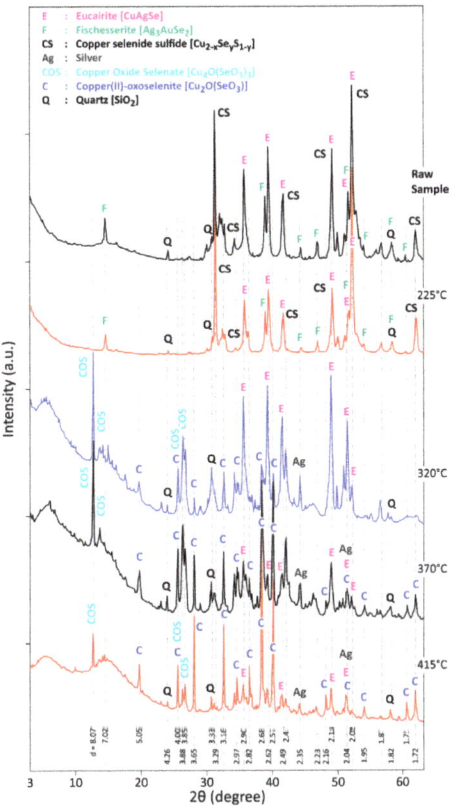

Figure 5. XRD patterns of CAS raw sample and the residues obtained during CAS sample treatment under air, from 225 °C to 415 °C.

At 505 °C (Figure 6), $[Cu_2O(SeO_3)]$ is still a main crystallized phase, but a new phase, tenorite (CuO) is also identified and becomes the predominant phase at 600 °C. Based on the %ML of Figure 4, on the results of XRD diffraction and data from the research work performed by Fokina et al. [36], the two steps of $Cu_4O(SeO_3)_3$ conversion into CuO and SeO_2 as final products can be described by Equations (2) and (3). Diffractograms of products obtained at 685 °C and 770 °C confirm the presence of well-crystallized phases of silver (gold), tenorite and quartz, which was present in all analyzed residues. Though some sulfur is present in the CAS sample, often substituted by selenium and/or tellurium, no solid phase containing sulfur was identified in the thermal treatment residue.

$$Cu_4O(SeO_3)_3 \rightarrow 2Cu_2O(SeO_3) + SeO_2 \tag{2}$$

$$2Cu_2O(SeO_3) \rightarrow 4CuO + 2SeO_2 \tag{3}$$

One may emphasize that no complete and systematic data of XRD analysis were found in the literature for monitoring the behavior of such copper anode slime under air in a wide temperature interval (225–770 °C). A comparison of XRD patterns of a CAS raw sample and its reaction product resulting from the thermal treatment of CAS at 770 °C in air is given in Figure 7. In summary, with a complex phase composition of a raw CAS sample (Eucairite-CuAgSe; copper selenide sulphide-$Cu_{2-x}Se_yS_{1-y}$; Fischesserite-Ag_3AuSe_2 and Quartz-SiO_2), the treatment residue at 770 °C is free of selenium bearing compounds and it is composed of cupric oxide, quartz and silver (gold). These compounds are concentrated in the residue as the thermal treatment allowed the volatilization of selenium dioxide, leading to a mass loss of sample exceeding 25 percent.

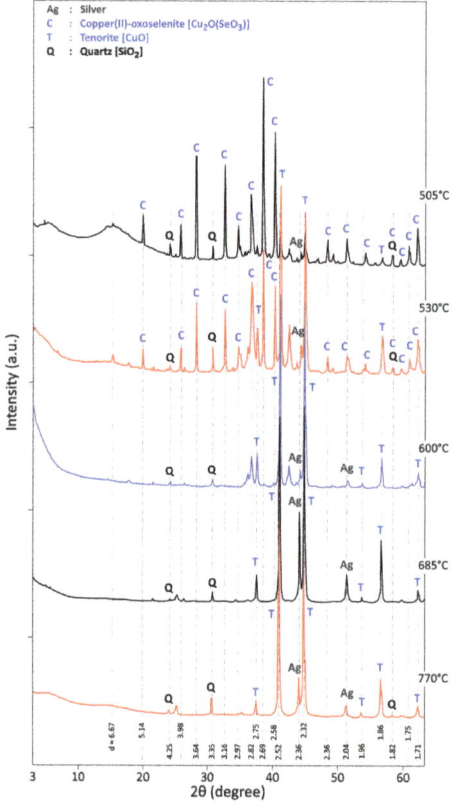

Figure 6. XRD patterns of the residues obtained during CAS sample treatment under air from 505 °C to 770 °C.

Likewise, several residues were examined by SEM-EDS technique. A typical morphology is given in Figure 8 with a visible difference in contrast suggesting the presence of areas of different compositions. Data of EDX analysis of spots 1 through 4 are summarized in Table 2. Within the measurement errors, the particle n° 1 is essentially composed of CuO (tenorite). The EDS analysis of spot n° 2 showed that silver (c.a., 78.6 wt %) is the major constituent with some gold (c.a., 14.6 wt %) and copper (c.a., 4.6 wt %). A possible zonal fusion may explain the smoothed shape of this particle. Copper and tellurium oxide (TeO_2) seem to be the main components of the area noted as n° 3. The composition of crystal-like n° 4 is similar of spot n° 2 with a quite different morphology. Note that selenium is not detected in the treatment residue. These data seem to confirm those obtained by XRD diffraction analysis.

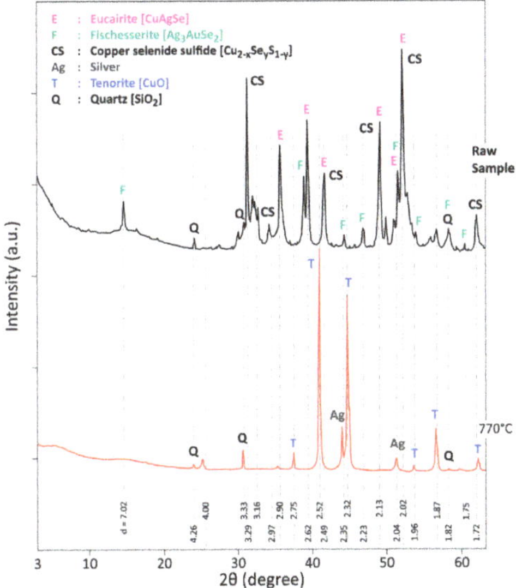

Figure 7. XRD patterns of a CAS raw sample and its treatment product in air at 770 °C.

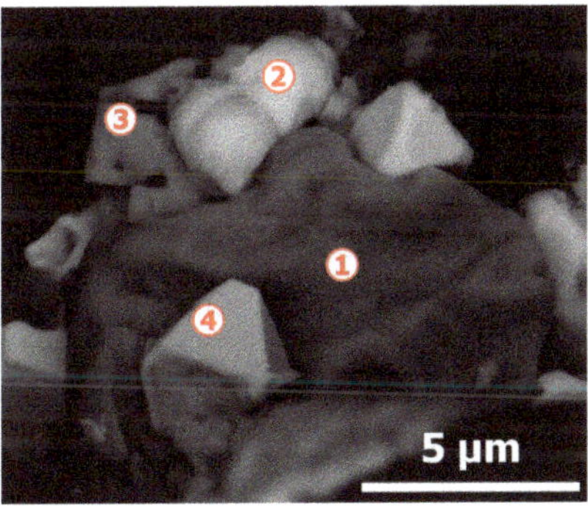

Figure 8. SEM aspects (backscattered electron micrograph) of the CAS sample treated at 770 °C in air.

Table 2. Elemental composition (EDS data) of the CAS treated at 770 °C in air atmosphere.

Elements	Spot n° 1		Spot n° 2		Spot n° 3		Spot n° 4	
	wt % [1]	at % [1]	wt %	at %	wt %	at %	wt %	at %
O	18.35	46.66	1.48	9.34	15.35	47.05	-	-
Al	0.59	0.89	-	-	0.67	1.22	-	-
Si	0.67	0.97	0.72	2.57	1.13	1.97	-	-
Cu	80.39	51.48	4.56	7.22	45.96	35.47	4.38	7.80
Ag	-	-	78.61	73.39	1.57	0.71	78.56	82.40
Au	-	-	14.62	7.48	-	-	17.06	9.80
Te	-	-	-	-	35.33	13.58	-	-

[1] w% and at % represents mass and atomic percentage, respectively.

During the treatment of CAS at temperatures above 500 °C, a solid mass is condensed in the tepid inner side of the experimental reactor. The visually observed quantity of this solid increased with the reaction progress and the rising temperature. A visible microscopy image of the recovered condensate is given in Figure 9 where centimetric needle-like crystals are observed. The SEM-EDS examination

(Figure 10) showed that the solid is homogenous, nevertheless some small zones of spherical and other shapes characterized by micronic sizes were revealed. As shown by Figure 10b, the matrix area of SEM image (Figure 10a) is composed especially of Se and O in atomic proportion close to the SeO_2 stoichiometry. In contrast, the lighter areas (Spot n° 2 in micrograph of Figure 10a) have a high content of selenium. Reduction of SeO_2 to Se° is probably achieved by sulfur dioxide (SO_2) during cooling of the gas phase. Taking into account that the crystallized phases synthesized during the oxidation of CAS by air (Figures 5 and 6) were [$Cu_4O(SeO_3)_3$] and [$Cu_2O(SeO_3)$], the selenium dioxide (SeO_2) which is generated from the decomposition of these compounds according to Equations (2) and (3) can be separated by volatilization and recovered in solid state.

Figure 9. Visible microscopy image of the solid condensate obtained during CAS treatment at 685°C.

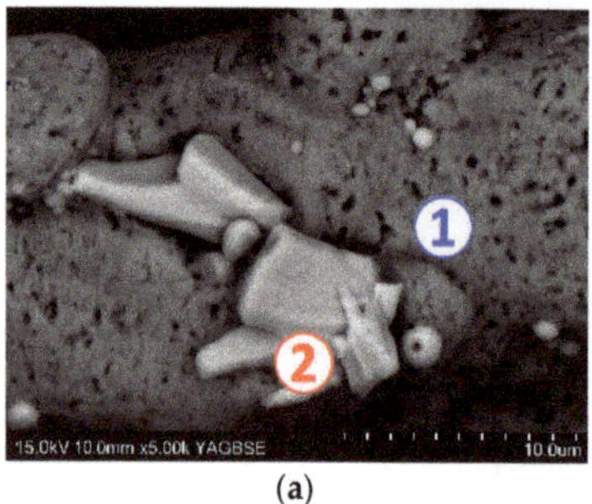

(a)

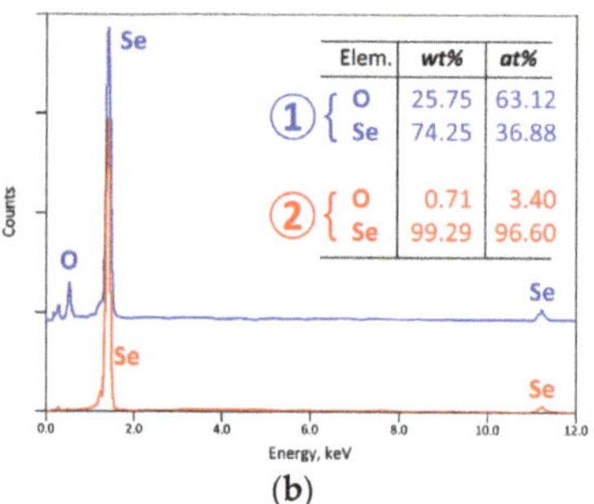

	Elem.	wt%	at%
① {	O	25.75	63.12
	Se	74.25	36.88
② {	O	0.71	3.40
	Se	99.29	96.60

(b)

Figure 10. SEM-EDS results of a condensate obtained during thermal treatment of the Cu-anode slime sample: (**a**) General view (backscattered electron micrograph) of SEM with two different phases; (**b**) EDS analysis of the same phases.

This study encompassed some characteristics for the thermal treatment of a copper by-product and allows understanding of the behavior of selected element compounds during the process. However, the end-of-life materials (e-wastes, thin films of used photovoltaics, spent catalysts, wasted flat screen …) containing rare, critical, strategic and high value elements present also a drawback, which is the presence of plastic materials. In addition, these plastics contain a considerable amount of halogenated substances based on chlorine (polyvinyl chloride-PVC) and bromine (Br-flame retardants combined with based Sb-synergist). The thermodynamic and kinetic reactivity of these substances with respect to targeted metals during an envisaged thermal process will be the subject of the second part of this investigation.

4. Conclusions

With grown in demand of current and future development of new technologies for rare, critical and strategic elements and facing the depletion of basic primary resources, the targeted element extraction and recycling from by-products, wastes and end-of-life materials is a promising choice.

Thermal treatment under an oxidizing atmosphere up to 770 °C of the copper anode slime (CAS), a potential source of these elements, led us to reach the following conclusions:

Selenium is mostly found in the copper-silver bearing phase (CuAgSe) and/or bounded to copper. Tellurium and sulfur can substitute selenium in the phase formulation. Gold is found in finest particles and Ag_3AuSe_2 seems to be the most likely gold bearing phase.

The thermal treatment of CAS at low temperatures led to the formation of double Cu-Se oxide $[Cu_4O(SeO_3)_3]$ which decomposed at temperatures higher than 400 °C.

Copper(II)-oxoselenite $[Cu_2O(SeO_3)]$, synthetized also at low temperatures, is stable, at least up to 530 °C. At higher temperatures, only tenorite (CuO) is found in the treatment residues as a decomposition product of these double oxides, while selenium dioxide is volatilized.

Almost pure SeO_2 is recovered by condensation of outlet gases issued from thermal treatment of CAS at temperatures higher than 500 °C.

The final residue of the CAS treatment under air beyond 600 °C is mostly composed of CuO and alloys of Ag-Au (with some Cu°), while tellurium is found as oxide (TeO_2), and silica (SiO_2) remains intact. Further treatment of such material is required for the final separation of these high value element compounds.

Author Contributions: Conceptualization, N.K., E.A., F.P. and J.Y.; Formal analysis, S.S., S.D. and F.D.; Investigation, N.K., S.D. and F.D.; Visualization, S.S. and F.D.; Resources, N.K., F.P. and J.Y.; Writing—original draft, N.K., E.A., S.S. and F.P.; Writing—review and editing, N.K., E.A., F.P. and J.Y.

Funding: This research was carried out within the framework of French projects namely: (i)—PEPS-Urban mine-2014 (CNRS-INSIS); (ii)—"Investissements d'avenir"—ANR-10-LABX-21-01/LABEX RESSOURCES21; (iii)—ANR CARNOT programme.

Conflicts of Interest: The authors declare no conflict of interest.

References

1. Available online: http://ec.europa.eu/growth/sectors/raw-materials/specific-interest/critical_en (accessed on 15 March 2019).
2. Allain, E.; Kanari, N.; Diot, F.; Yvon, J. Development of a process for the concentration of the strategic tantalum and niobium oxides from tin slags. *Miner. Eng.* **2019**, *134*, 97–103. [CrossRef]
3. Holgersson, S.; Steenari, D.-M.; Björkman, M.; Cullbrand, K. Analysis of the metal content of small-size Waste Electric and Electronic Equipment (WEEE) printed circuit boards—Part 1: Internet routers, mobile phones and smartphones. *Resour. Conserv. Recycl.* **2018**, *133*, 300–308. [CrossRef]
4. Kumar, S.; Rawat, S. Future e-Waste: Standardisation for reliable assessment. *Gov. Inf. Q.* **2018**, *35*, S33–S42. [CrossRef]
5. Otto, S.; Kibbe, A.; Henn, L.; Hentschke, L.; Kaiser, F.G. The economy of E-waste collection at the individual level: A practice oriented approach of categorizing determinants of E-waste collection into behavioral costs and motivation. *J. Clean. Prod.* **2018**, *204*, 33–40. [CrossRef]
6. Jayaraman, K.; Vejayon, S.; Raman, S.; Mostafiz, I. The proposed e-waste management model from the conviction of individual laptop disposal practices-An empirical study in Malaysia. *J. Clean. Prod.* **2019**, *208*, 688–696. [CrossRef]
7. Abbondanza, M.; Souza, R. Estimating the generation of household e-waste in municipalities using primary data from surveys: A case study of Sao Jose dos Campos, Brazil. *Waste Manag.* **2019**, *85*, 374–384. [CrossRef]
8. Cho, B.-G.; Lee, J.-C.; Yoo, K. Valuable Metal Recycling. *Metals* **2018**, *8*, 345. [CrossRef]
9. Ding, Y.; Zhang, S.; Liu, B.; Zheng, H.; Chang, C.-C.; Ekberg, C. Recovery of precious metals from electronic waste and spent catalysts: A review. *Resour. Conserv. Recycl.* **2019**, *141*, 284–298. [CrossRef]
10. Avarmaa, K.; Klemettinen, L.; O'Brien, H.; Taskinen, P. Urban mining of precious metals via oxidizing copper smelting. *Miner. Eng.* **2019**, *133*, 95–102. [CrossRef]
11. Amer, A. Processing of copper anodic-slimes for extraction of valuable metals. *Waste Manag.* **2003**, *23*, 763–770. [CrossRef]

12. Khaleghi, A.; Ghader, S.; Afzali, D. Ag recovery from copper anode slime by acid leaching at atmospheric pressure to synthesize silver nanoparticles. *Int. J. Min. Sci. Technol.* **2014**, *24*, 251–257. [CrossRef]

13. Li, X.J.; Yang, H.Y.; Jin, Z.N.; Chen, G.B.; Tong, L.L. Transformation of Selenium-Containing Phases in Copper Anode Slimes During Leaching. *JOM* **2017**, *69*, 1932–1938. [CrossRef]

14. Xiao, L.; Wang, Y.; Yu, Y.; Fu, G.; Han, P.; Sun, Z.; Ye, S. An environmentally friendly process to selectively recover silver from copper anode slime. *J. Clean. Prod.* **2018**, *187*, 708–716. [CrossRef]

15. Kilic, Y.; Kartal, G.; Timur, S. An investigation of copper and selenium recovery from copper anode slimes. *Int. J. Miner. Process.* **2013**, *124*, 75–82. [CrossRef]

16. Gaballah, I.; Allain, E.; Djona, M. Extraction of tantalum and niobium from tin slags by chlorination and carbochlorination. *Metall. Mater. Trans. B* **1997**, *28*, 359–369. [CrossRef]

17. Kanari, N.; Gaballah, I.; Allain, E. Kinetics of oxychlorination of chromite: Part I. Effect of temperature. *Thermochim. Acta* **2001**, *371*, 143–154. [CrossRef]

18. Kanari, N.; Gaballah, I.; Allain, E. Kinetics of oxychlorination of chromite: Part II. Effect of reactive gases. *Thermochim. Acta* **2001**, *371*, 75–86. [CrossRef]

19. Kanari, N.; Allain, E.; Joussemet, R.; Mochón, J.; Ruiz-Bustinza, I.; Gaballah, I. An overview study of chlorination reactions applied to the primary extraction and recycling of metals and to the synthesis of new reagents. *Thermochim. Acta* **2009**, *495*, 42–50. [CrossRef]

20. Kanari, N.; Menad, N.-E.; Ostrosi, E.; Shallari, S.; Diot, F.; Allain, E.; Yvon, J. Thermal Behavior of Hydrated Iron Sulfate in Various Atmospheres. *Metals* **2018**, *8*, 1084. [CrossRef]

21. Kanari, N.; Evrard, O.; Neveux, N.; Ninane, L. Recycling ferrous sulfate via super-oxidant synthesis. *JOM* **2001**, *53*, 32–33. [CrossRef]

22. Kanari, N.; Ostrosi, E.; Ninane, L.; Neveux, N.; Evrard, O. Synthesizing alkali ferrates using a waste as a raw material. *JOM* **2005**, *57*, 39–42. [CrossRef]

23. Kanari, N.; Filippov, L.; Diot, F.; Mochon, J.; Ruiz-Bustinza, I.; Allain, E.; Yvon, J. Synthesis of potassium ferrate using residual ferrous sulfate as iron bearing material. *J. Phys. Conf. Ser.* **2013**, *416*, 012013. [CrossRef]

24. Kanari, N.; Filippova, I.; Diot, F.; Mochón, J.; Ruiz-Bustinza, I.; Allain, E.; Yvon, J. Utilization of a waste from titanium oxide industry for the synthesis of sodium ferrate by gas–solid reactions. *Thermochim. Acta* **2014**, *575*, 219–225. [CrossRef]

25. Zhang, L.; Xu, Z. A critical review of material flow, recycling technologies, challenges and future strategy for scattered metals from minerals to wastes. *J. Clean. Prod.* **2018**, *202*, 1001–1025. [CrossRef]

26. Davidsson, S.; Höök, M. Material requirements and availability for multi-terawatt deployment of photovoltaics. *Energy Policy* **2017**, *108*, 574–582. [CrossRef]

27. Padoan, F.C.; Altimari, P.; Pagnanelli, F. Recycling of end of life photovoltaic panels: A chemical prospective on process development. *Sol. Energy* **2019**, *177*, 746–761. [CrossRef]

28. Domínguez, A.; Geyer, R. Photovoltaic waste assessment of major photovoltaic installations in the United States of America. *Renew. Energy* **2019**, *133*, 1188–1200. [CrossRef]

29. Chen, T.T.; Dutrizac, J.E. A mineralogical study of the deportment and reaction of silver during copper electrorefining. *Metall. Mater. Trans. B* **1989**, *20*, 345–361. [CrossRef]

30. Petkova, E. Microscopic examination of copper electrorefining slimes. *Hydrometallurgy* **1990**, *24*, 351–359. [CrossRef]

31. Petkova, E. Hypothesis about the origin of copper electrorefining slime. *Hydrometallurgy* **1994**, *34*, 343–358. [CrossRef]

32. Chen, T.T.; Dutrizac, J.E. Mineralogical characterization of a copper anode and the anode slimes from the la caridad copper refinery of mexicana de cobre. *Metall. Mater. Trans. B* **2005**, *36*, 229–240. [CrossRef]

33. Available online: http://s1.iran-mavad.com/ASM%20hanbooks/Vol_3_ASM%20handbooks_iran-mavad.com.pdf (accessed on 29 March 2019).

34. Roine, A. *Outokumpu HSC Chemistry for Windows*; Version 3.0; Outokumpu Research: Pori, Finland, 1997.

35. Portnichenko, P.Y.; Romhanyi, J.; Onykiienko, Y.A.; Henschel, A.; Schmidt, M.; Cameron, A.S.; Surmach, M.A.; Lim, J.A.; Park, J.T.; Schneidewind, A.; et al. Magnon spectrum of the helimagnetic insulator Cu_2OSeO_3. *Nat. Commun.* **2016**, *7*, 10725. [CrossRef] [PubMed]
36. Fokina, E.L.; Klimova, E.V.; Charykova, M.V.; Krivovichev, V.G.; Platonova, N.V.; Semenova, V.V.; Depmeier, W. The thermodynamics of arsenates, selenites, and sulfates in the oxidation zone of sulfide ores: VIII. Field of thermal stability of synthetic analog of chalcomenite, its dehydration and dissociation. *Geol. Ore Deposits* **2014**, *56*, 538–545. [CrossRef]

© 2019 by the authors. Licensee MDPI, Basel, Switzerland. This article is an open access article distributed under the terms and conditions of the Creative Commons Attribution (CC BY) license (http://creativecommons.org/licenses/by/4.0/).

Article

Thermochemical Route for Extraction and Recycling of Critical, Strategic and High-Value Elements from By-Products and End-of-Life Materials, Part II: Processing in Presence of Halogenated Atmosphere

Ndue Kanari [1,*], Eric Allain [1], Seit Shallari [2], Frédéric Diot [1], Sébastien Diliberto [3], Fabrice Patisson [3] and Jacques Yvon [1]

[1] CNRS, GeoRessources, Université de Lorraine, F-54000 Nancy, France; ericgallain@gmail.com (E.A.); frederic.diot@univ-lorraine.fr (F.D.); jyvon6355@gmail.com (J.Y.)

[2] Faculty of Agriculture and Environment, Agricultural University of Tirana, 1029 Tirana, Albania; seitshallari@gmail.com

[3] CNRS, Labex DAMAS, IJL, Université de Lorraine, F-54000 Nancy, France; sebastien.diliberto@univ-lorraine.fr (S.D.); fabrice.patisson@univ-lorraine.fr (F.P.)

* Correspondence: ndue.kanari@univ-lorraine.fr; Tel.: +33-372-744-530

Received: 16 July 2020; Accepted: 18 September 2020; Published: 21 September 2020

Abstract: During the treatment of copper anode slime (CAS) under an air atmosphere, several aspects of the interactions of its main components (CuAgSe, $Cu_{2-x}Se_yS_{1-y}$, Ag_3AuSe_2) with oxygen were described in Part I. As a comparative and complementary study, this work deals with the thermal behavior of CAS under air in the presence of polyvinyl chloride (PVC) between 195 and 770 °C. The preliminary thermal treatment of an e-waste sample containing brominated substances was also performed. The reaction products were systematically analyzed by scanning electron microscopy through energy-dispersive spectroscopy (SEM-EDS) and X-ray diffraction (XRD) to investigate the thermal behaviors of the studied samples in a halogenated medium. At low temperatures, the copper, silver and selenium compounds of the CAS reacted with the HCl, issued from PVC degradation, leading to the formation of their respective chlorides. Bromides of valuable metals (Cu, Pb, Sn . . .) were synthesized during the e-waste treatment at 500 °C and they were distributed between the solid residue and gaseous phase. The data obtained give an insight into the reactivity of several metals towards halogenated substances, which may be valuable information for conducting the extraction and recycling of targeted elements from industrial by-products and end-of-life materials by a thermochemical route.

Keywords: critical and strategic materials; valuable metals; copper anode slime; PVC; de-chlorination; e-waste; thermal treatment; halogenation; metal halide volatilization

1. Introduction

The results described in Part I [1] of this investigation combined with research works cited therein [2–16] gave a general view of needs and secondary sources for a large range of materials, identified frequently as critical and strategic materials, and they are indispensable for present and forthcoming innovations in renewable energies, transportation and cutting-edge technologies.

These materials, mostly metals, are often concentrated in the by- and/or co-products of base metals extraction and they are subsequently separated and recovered employing pyro-and/or hydrometallurgical processes. It is expected that the actual and future demand for several of these materials will grow strongly, especially to feed the production of electric and electronic devices, while the lifespans of many products such as computers and cell phones are decreasing rapidly [4].

Materials **2020**, *13*, 4203; doi:10.3390/ma13184203   www.mdpi.com/journal/materials

This trend, among others, leads to the generation of a huge waste stream (electrical waste and electronic equipment, WEEE), often designated as e-waste [5–8]. Its elevated content in high added value elements (precious, strategic, critical, rare and rare earths elements), particularly for "smart parts", makes this waste stream, a very attractive "urban mine" for the extraction of these components. But, it is necessary to emphasize that the hidden side of the e-waste stream is not all golden. These materials are composed on average of about 35% of metals, 35% of refractory oxides and ceramics and 30% of various plastics and resins. In addition to the complexity and miniaturization of the targeted metals which make their recovery difficult, the presence of chlorinated and bromated substances associated with plastics and many other components poses serious environmental problems during the recycling of e-waste. With appreciated functional properties, one of these plastics, polyvinyl chloride (PVC) is used in a broad range of products including electric and electronic appliances. With about 57% Cl (in pure PVC), it is an inherently fire-resistant plastic. However, the end of life of PVC is a real problematic case due to its high chlorine and diverse additives content. Other sources of the halogen-containing flame retardants based especially on bromine substances are widely used to meet the flammability standards of products. Several recent scientific works dealing with diverse aspects of plastics containing PVC and halogenated flame retardants are well summarized in Materials [17–22].

There are not many adequate solutions to e-waste treatment and the recycling of the contained metals. One may mention the "hot process" which resembles incineration and / or conventional pyro-metallurgical processes for the extraction of base metals (Cu, Pb ...). The presence of commonly halogenated plastics complicates the operation at high temperature and causes the loss of some valuable elements. The "wet process", meanwhile, relies on multiple steps, apparatuses and the use of a wide range of inorganic and organic reagents which are not always environmentally friendly. The use of pyrometallurgical and hydrometallurgical processes for the recovery of high-value metals from wastes was reviewed by Ding et al. [10]. Regarding smelting, it was noted that it is beneficial for the concentration of precious metals. Avarmaa et al. [11] indicated that several precious metals (gold, silver, palladium and platinum), found in end-of-life electronics, are recovered in the melted copper using a smelting process. It should be noted that high-value elements (Au, Ag, Se, Te, platinum group metals, etc.) are concentrated in the anode slime of primary copper extraction from sulfide materials. One may conclude that the processing of electronic wastes in conventional copper pyrometallurgy resulted in the concentration of the valuable elements in by-products such as the anode slime of the copper electro-refinery.

Recent studies dealt with various treatments of copper anode slime (CAS) recovering targeted elements. Among various methods used, the treatment by H_2SO_4-O_2 and thiourea was given by Amer [12]. Leaching processes of the CAS using H_2SO_4 and HNO_3 as digestion agents were also applied [13,14]. A combined approach, thiosulfate leaching followed by an electrodeposition, was described by Xiao et al. [15]. Use of NaOH for selenium dissolution was also developed [16]. In Part I of this investigation, the thermal behavior of CAS under an oxidizing atmosphere was described. Of special emphasis during this treatment was the recuperation of pure selenium oxide (SeO_2) by cooling the gaseous phase during the CAS treatment.

However, limited current works are available for the thermal treatment of CAS under a dry halogenated atmosphere. In this framework, the work presented here deals with the thermal behavior of CAS supplied by a European copper plant under air in presence of polyvinyl chloride (PVC). A systematic physico-chemical characterization of the treatment products obtained under different experimental conditions has been developed in order to understand the reaction paths and mechanisms of the process for achieving a selective separation of the elements contained in the CAS. A comparison of results obtained during the treatment of CAS in solely air atmosphere with those obtained in air + PVC is often exhibited to understand the intermediate steps involved during thermochemical treatment of the CAS. Furthermore, the preliminary thermal treatment of an e-waste sample containing brominated substances was also performed. This study was performed in the frame our research program focusing mostly on the treatment of low-grade raw materials and solid residues by a thermo-chemical route [23–28] and on the valorization of certain iron bearing wastes via chemical synthesis of new and green materials [29–34].

2. Materials and Methods

A sample of CAS, with a median particles size lower than 63 μm, was used for this research work. Diverse analytical methods were used for the physico-chemical characterization of the CAS sample, but only the results of scanning electron microscopy–energy-dispersive spectroscopy (SEM-EDS) and X-ray diffraction (XRD) analyses are shown in this paper. The SEM-EDS and XRD equipment as well as the analytic protocol were described in Part I [1].

A sample of pure polyvinyl chloride (PVC) (without additives) was also used for this study. The e-waste sample was constituted of printed circuit boards (PCBs) obtained from end-of-life computers.

The PVC sample and the mixture of (CAS + PVC) sample with a mass ratio CAS/PVC = 1 were conditioned as pellets of approximately centimetric size (Figure 1) prior to thermal treatment. The PCBs were cut into strips sectioned at (10 × 2) cm without further treatment. Isothermal experimental tests for the treatment of CAS, PVC samples and mixtures of CAS with PVC (CAS + PVC) as well as for the e-waste sample were conducted in horizontal system including air metering and a static tubular furnace able to reach 1600 °C with a uniform temperature segment of at least 20 cm. The specimen holder (Figure 1) and the reactor were made of quartz material which is resistant towards the working conditions. To experiment with this process, a pre-weighted sample of several grams was introduced directly into the furnace preheated at a fixed temperature. When the dwell time is reached, the sample is removed from the furnace and cooled down to room temperature. Condensation of the outlet gases at room temperature led to the recovery of the vapor phase as condensates. Solid products (residues and condensates) were examined by SEM-EDS and XRD analysis.

Figure 1. Optical image of polyvinyl chloride (PVC) and (copper anode slime (CAS + PVC) pellets used for the thermal treatment.

3. Results

3.1. Elemental and Mineralogical Analysis of Copper Anode Slime (CAS) and Polyvinyl Chloride (PVC) Samples

As noted in Part I [1], several research reports [12–16,35–38] emphasized the complex nature of the CAS based on their elemental, mineralogical and morphological analysis. The particular morphology of the studied CAS sample is also exhibited by an SEM image as illustrated in Figure 2a, meanwhile the multi-elemental composition (Cu, Se, Ag, S, O, Au, Si, Te, Cl) is clear in the general EDS spectrum of the CAS (Figure 2b). The presence of carbon in the EDS spectrum is probably resulting from the carbonaceous matter in the CAS sample and the carbon coating used to make the sample conductive for the SEM-EDS analysis.

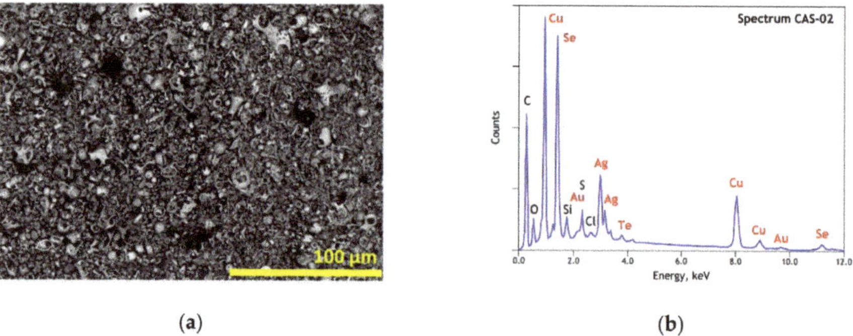

Figure 2. Scanning electron microscopy–energy-dispersive spectroscopy (SEM-EDS) results of initial copper anode slime: (**a**) general view (backscattered electron micrograph "BSE") of the used sample; (**b**) overall EDS analysis of the used sample.

A clearer view revealing the atypical morphology of the CAS sample is shown in Figure 3 and, as depicted previously [1], there are multiple particles that are irregularly and shell shaped. The SEM-EDS microanalysis data given in Table 1 indicates Cu, Se and Ag (spot n° 1, 4 and 5) are frequently associated. Sulfur and to some level tellurium are also found in the spots probed; finally, gold is identified frequently in the finest particles (spot n° 4 and 5). Several explanations for the particular state of CAS particles and their composition are provided in Part I [I] of this investigation and there are related mostly with the thermochemical reactions developing during smelting and refinery of copper.

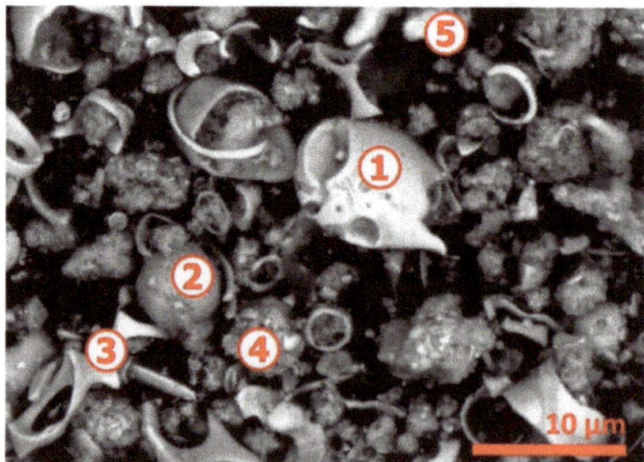

Figure 3. SEM-BSE image of initial copper anode slime. Numbers 1 to 5 indicate the spots for microanalysis.

Table 1. Elemental composition of copper anode slime analyzed by SEM-EDS.

Elements	Spot $n°$ 1		Spot $n°$ 2		Spot $n°$ 3		Spot $n°$ 4		Spot $n°$ 5	
	[1]wt%	[1]at%	wt%	at%	wt%	at%	wt%	at%	wt%	at%
O	-	-	-	-	-	-	1.75	8.15	3.49	15.70
Si	-	-	-	-	-	-	0.31	0.82	0.49	1.25
S	1.17	2.87	3.65	7.84	2.05	4.44	3.30	7.66	2.06	4.62
Cu	32.40	40.13	47.16	51.11	48.48	53.10	32.60	38.22	28.05	31.73
Se	33.15	33.05	41.97	36.60	44.68	39.37	28.99	27.34	23.06	20.99
Ag	30.27	22.09	5.50	3.51	4.79	3.09	17.01	11.74	33.38	22.24
Te	3.01	1.86	1.71	0.92	-	-	-	-	-	-
Au	-	-	-	-			16.04	6.07	9.47	3.46

[1] wt% and at% represent mass and atomic percentage, respectively.

The XRD diffractogram of CAS is drawn in Figure 4. It detects the presence of eucairite (CuAgSe), confirming the microanalysis performed by SEM-EDS. Furthermore, there are, in the diffractogram of CAS, some peaks corresponding to Cu_2Se, Cu_2S, $CuSeS$, as well as to non-stoichiometric compounds with an absence of certain peaks probably due to crystallite orientation and these phases are defined as copper selenide sulfide ($Cu_{2-x}Se_yS_{1-y}$). The XRD peaks for the Au-bearing phase match mostly with fischesserite (Ag_3AuSe_2). This phase composition (with substitution of Ag by Cu) is also revealed by SEM-EDS analysis (spot n° 4 and 5 in Table 1). Finally, Quartz (SiO_2) completes the list of crystallized phases identified by XRD in the CAS sample.

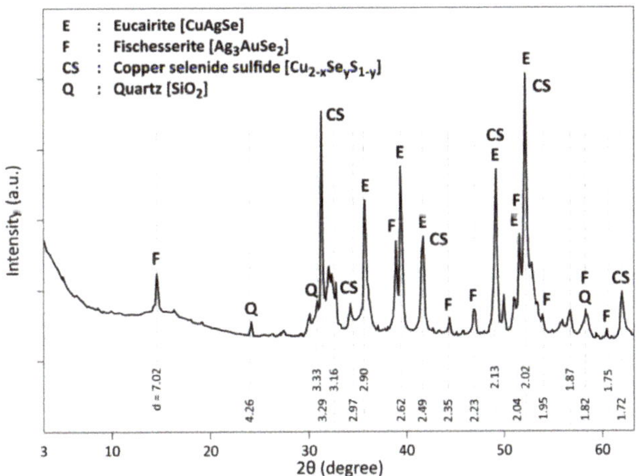

Figure 4. X-ray diffractogram (XRD) of copper anode sample; data adapted from [1].

The pure PVC sample is constituted of near-spherical particles shapes of size less than 100 µm (Figure 5a). As expected, chlorine and carbon are the only elements revealed (Figure 5b) within the element detection threshold of the SEM-EDS instrument. The XRD pattern of the used PVC sample, exhibited several diffuse halos indicating the amorphous nature of PVC.

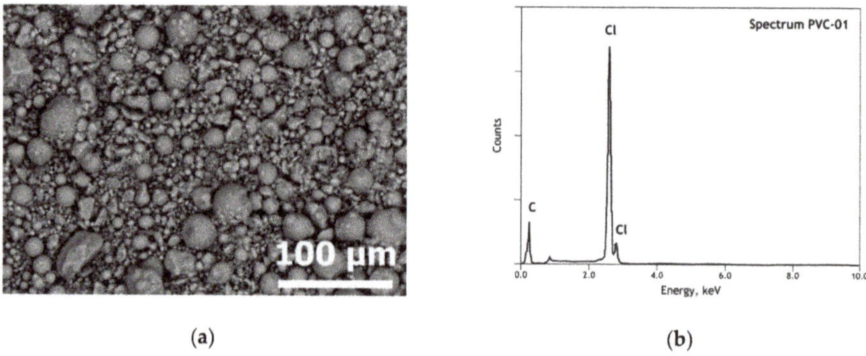

(a)   (b)

Figure 5. SEM-EDS results of initial PVC: (**a**) general view (BSE micrograph) of the used sample; (**b**) overall EDS analysis of the used sample.

3.2. Thermal Treatment of a Mixture of (CAS + PVC) and PVC Sample in Air for 1 h

Several grams of the copper anode sample mixed with PVC (CAS + PVC, CAS/PVC = 1) and conditioned as pellets (Figure 1) were used for isothermal tests performed between 195 and 770 °C in air atmosphere with a flow rate of 25 L/h. The evolution of the % mass loss (%ML) as a function of the treatment temperature is plotted in Figure 6. The %ML curve shape presents a regular continuous mass loss rise for temperatures up to about 600 °C. The examination of Figure 6 with the data obtained for the treatment of CAS alone in air atmosphere (Figure 7) shows a big difference in the curve shape, at least at low temperature. The mass gain in Figure 7 was attributed to the oxidation of several CAS elements and synthesis of combined oxides [$Cu_4O(SeO_3)_3$, $Cu_2O(SeO_3)$], thereafter, their decomposition and subsequent volatilization of SeO_2 led to the mass loss observed at higher temperature [1].

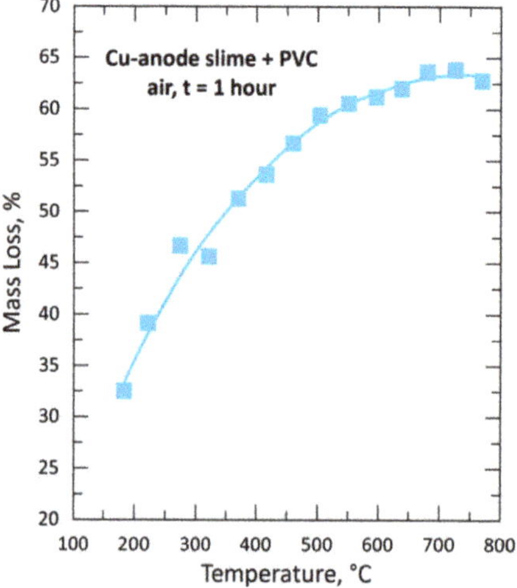

Figure 6. Plot of the mass loss of the sample versus temperature during treatment of CAS + PVC in air for 1 h.

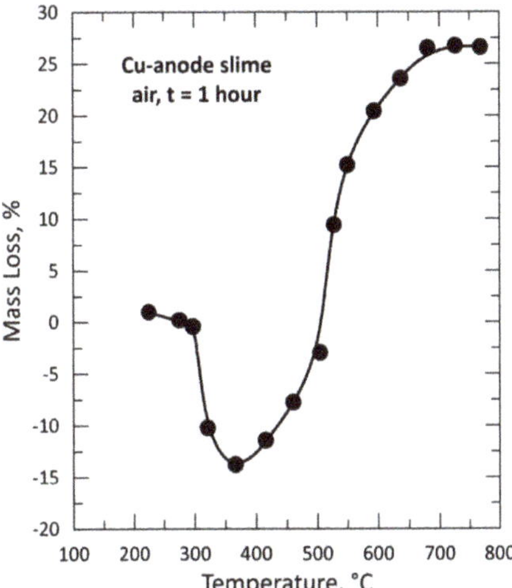

Figure 7. Plot of the mass loss of the sample versus temperature during treatment of CAS in air for 1 h.

One option to clarify these different behaviors was to check the reaction of PVC with air at various temperatures. Additives and composition information are identified by analyzing the reaction products of the thermal treatment of (CAS + PVC) in air atmosphere, as described in Section 3.3.

Figure 8 is a plot of %ML of the PVC sample versus temperature ranging from 165 to 505 °C. It is clear from this figure that there are two distinguishable steps describing the thermal treatment of PVC in air atmosphere. The first step (T < 300 °C) could be attributed to a complete removal of chlorine during the thermal degradation of PVC. The value of 56.62 % ML (corresponding to the theoretical chlorine content of PVC) at 275 °C seems to indicate that all chlorine is volatilized (as HCl) leaving behind a carboniferous residue (char). The removal of chlorine is also confirmed by SEM-EDS analysis of the residue obtained at 275 °C. Other confirmation of HCl generation was also the acidity (pH) monitoring of an alkaline solution scrubbing out-gases issued from PVC treatment between 225 and 275 °C. After a latent period, the solution pH decreased abruptly from 12.5 to 0.50 as the reaction progress at 250 °C. As shown by the Arrhenius diagram in the miniaturized graphic in Figure 8, the mean apparent activation energy (E_a) for the chlorine removal is calculated to be about 120 kJ/mol between 165 and 250 °C. The second step for the treatment of PVC in air was observed beyond 300 °C and may attributed to the reaction of the hydrocarbons with oxygen giving carbon oxides and water vapor as a final reaction product.

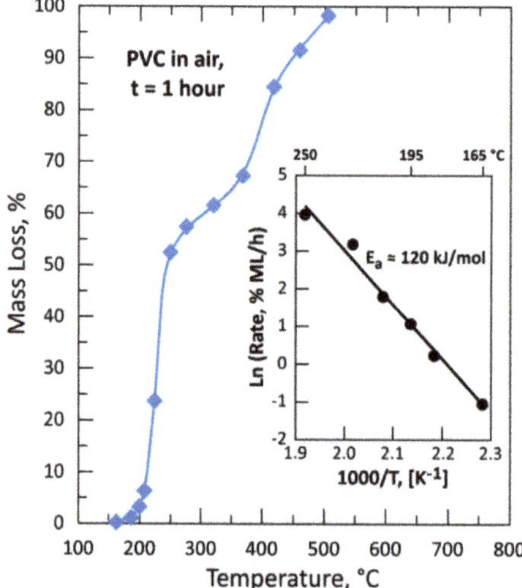

Figure 8. Thermal behavior of a pure PVC sample in air for 1 h.

3.3. Analysis of the Reaction Products

Residues produced from isothermal processing of CAS + PVC under air atmosphere were evaluated by XRD and SEM-EDS techniques and results were compared with those obtained during the treatment of only CAS in air atmosphere.

Figure 9 displays the XRD patterns of the residue of CAS+PVC treated at 195 °C. Besides the characteristic peaks of the initial CAS phases (CuAgSe, Ag_3AuSe_2, $Cu_{2-x}Se_yS_{1-y}$ and SiO_2), there are new crystallized phases, namely cuprous chloride (CuCl), silver chloride (AgCl), and umangite (Cu_3Se_2). With this phase identification and based on thermodynamic calculations [39], one can deduce that the following overall reactions may have occurred:

$$1/4Cu_2Se_{(s)} + HCl_{(g)} + 1/4O_{2(g)} \rightarrow 1/2CuCl_{(s)} + 1/4SeCl_{2(g)} + 1/2H_2O_{(g)} \tag{1}$$

$$2Cu_2Se_{(s)} + HCl_{(g)} + 1/4O_{2(g)} \rightarrow Cu_3Se_{2(s)} + CuCl_{(s)} + 1/2H_2O_{(g)} \tag{2}$$

$$1/4Ag_2Se_{(s)} + HCl_{(g)} + 1/4O_{2(g)} \rightarrow 1/2AgCl_{(s)} + 1/4SeCl_{2(g)} + 1/2H_2O_{(g)} \tag{3}$$

The standard Gibbs energy changes ($\Delta_r G°$) at 100 °C [39] for the reaction (1), (2) and (3) are −76.41, −116.19 and −68.84 kJ/mol HCl, respectively showing the thermodynamic reactivity of the targeted phases with respect to HCl in presence of oxygen. The selenium dichloride ($SeCl_2$) is considered for the thermodynamic calculation among the selenium chlorides (Se_2Cl_2, $SeCl_2$, $SeCl_4$, $SeOCl_2$), although it is difficult to give an exact reactional scheme of selenium chloride formation. However, the formation of CuCl (chlorine being in the PVC) instead of CuO and the volatilization of selenium chloride (s) explain the mass loss of sample (CAS + PVC) treated at low temperature (see Figure 6). The characteristic XRD peaks of Cu_3Se_2 is observed in the XRD patterns of (CAS + PVC) treated at 195 °C (Figure 9). The phase diagram of the system Cu-Se [40] indicates the presence of Cu_3Se_2 which melted incongruently over 112 °C. One may assume that the Cu_3Se_2 appeared (Equation (2)) as a stable phase, perhaps during the sample cooling.

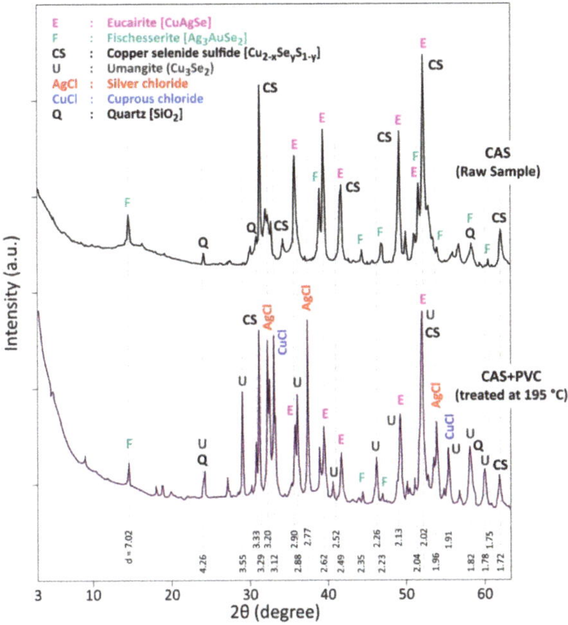

Figure 9. XRD patterns of a CAS raw sample and (CAS + PVC) treatment product in air at 195 °C.

The presence of the partially chlorinated CuAgSe phase is observed by SEM-EDS examination for a residue obtained at 225 °C. The spot n° 2 of Figure 10 seems to indicate a small particle likely to be the CuAgSe phase. Its neighbor (spot n° 1) should be a mixture between the CuAgSe phase and Cu-Ag chlorides that are in solid state at this temperature.

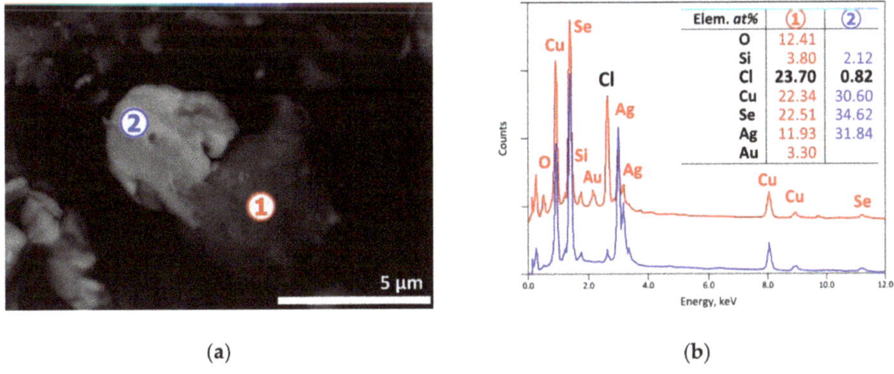

Elem. *at%*	(1)	(2)
O	12.41	
Si	3.80	2.12
Cl	23.70	0.82
Cu	22.34	30.60
Se	22.51	34.62
Ag	11.93	31.84
Au	3.30	

(a) (b)

Figure 10. SEM-EDS results of a residue issued from the (CAS+PVC) treatment at 225 °C in air atmosphere: (**a**) general view (BSE micrograph); (**b**) EDS analysis of spot n° 1 and 2.

The XRD results of the residues (CAS+PVC) obtained at temperatures higher than 320 °C show a good similarity with those obtained during the treatment of CAS in the absence of PVC. However, as shown in Figure 11, the decomposition of neo-formed combined oxides [e.g., Cu₄O(SeO₃)₃] and the appearance of new phases [e.g., CuO], occurred apparently, at lower temperatures in the case of CAS treatment in presence of PVC. One of the plausible hypotheses is that the temperature in the reaction zone can be significantly higher than the fixed furnace temperature due to the highly exothermic nature of the reactions of the PVC degradation under oxygen. Such exothermic phenomena were also observed previously during the treatment of only 2 g of sulfides with chlorine at 300 °C resulting in a temperature increase in the reaction zone of about 30 °C [41,42].

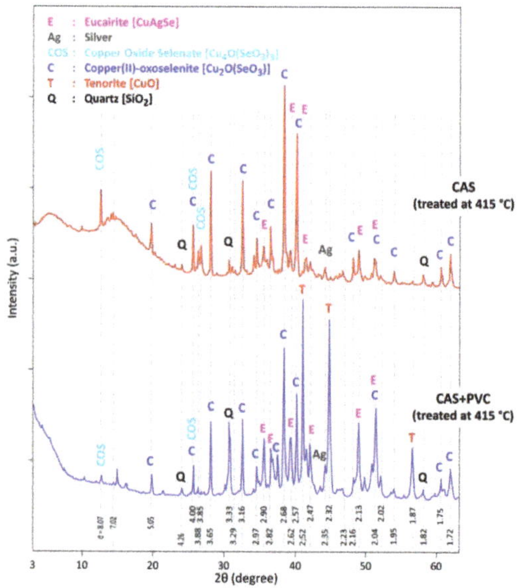

Figure 11. Comparison of XRD patterns of the product issued from the treatment of CAS and (CAS + PVC) samples in air at 415 °C.

A summary of the crystalline phases identified by XRD in the treatment residues obtained during the treatment of CAS and that of the mixture (CAS + PVC) in air for 1 h is given in Tables 2 and 3, respectively. As shown in Table 2, the phases of the residue obtained at 225 °C are those revealed in the CAS raw sample. Eucairite (CuAgSe) is still stable at 320 °C, while the characteristic peaks of $Cu_{2-x}Se_yS_{1-y}$ and Ag_3AuSe_2 phases disappeared at this temperature. The XRD of residue produced at 320 °C indicated the formation of new phases such as: [$Cu_4O(SeO_3)_3$], [$Cu_2O(SeO_3)$]. Metallic silver is also identified in this treatment residue; note that the main XRD peaks of Ag° and Au° are overlapped. The panorama of the residues phases at 415 °C is identical of that of 320 °C, though the intensities proportion of phases changed suggesting the appearance of new phases.

Table 2. Summary of XRD results of the residues issued from the treatment of CAS in air for 1 h.

Identified Phases	IS [1]	225 °C	320 °C	415 °C	505 °C	685 °C	770 °C
CuAgSe	■	■	■	■			
Ag_3AuSe_2	■	■					
$Cu_{2-x}Se_yS_{1-y}$	■	■					
Silver (Ag)			■	■	■	■	■
$Cu_4O(SeO_3)_3$			■	■			
$Cu_2O(SeO_3)$			■	■	■		
CuO						■	■
SiO_2	■	■	■	■	■	■	■

[1] Initial Sample.

Table 3. Summary of XRD results of the residues issued from the treatment of the mixture (CAS + PVC) in air for 1 h at various temperatures.

Identified Phases	IS [1]	195 °C	225 °C	320 °C	415 °C	505 °C	685 °C	770 °C
$CuAgSe$	●	●	●	●	●	●		
Ag_3AuSe_2	●	●	●					
$Cu_{2-x}Se_yS_{1-y}$	●	●	●					
Cu_3Se_2		●	●					
$CuCl$		●	●					
Silver (Ag)				●	●	●	●	●
$AgCl$		●	●					
$Cu_4O(SeO_3)_3$				●	●			
$Cu_2O(SeO_3)$				●	●	●		
CuO					●	●	●	●
SiO_2	●	●	●	●	●	●	●	●

[1] Initial sample.

At 505 °C [$Cu_2O(SeO_3)$] is still the main crystallized phase, but tenorite (CuO) is also identified and becomes the main phase at higher temperature. According to the data reported in Part I [1] and those given by Fokina et al. [43], the two steps of $Cu_4O(SeO_3)_3$ conversion into CuO and SeO_2 as final products can be described by Equations (4) and (5):

$$Cu_4O(SeO_3)_3 \rightarrow 2Cu_2O(SeO_3) + SeO_2 \tag{4}$$

$$2Cu_2O(SeO_3) \rightarrow 4CuO + 2SeO_2 \tag{5}$$

Analysis of the XRD diffractograms for the residues obtained at 685 °C and 770 °C confirms the presence of well-crystallized phases of silver (Ag°), tenorite (CuO) and quartz (SiO_2), which was present in all analyzed residues.

The reaction of HCl released during the partial degradation of the PVC with CAS at 195 °C (Table 3) led to the synthesis of new crystallized phases such as: $CuCl$, Cu_3Se_2 and $AgCl$ which are still identified in the treatment residue obtained at 225 °C because at this temperature the rate of de-chlorination of PVC is slow and continuously producing the necessary HCl for 1 h of the experimental test (see Figure 8). At 320 °C, the phases identified in the treatment residues of CAS in the absence (Table 2) or in the presence of PVC (Table 3) are similar. One may hypothesize that starting from this temperature treatment, the kinetic of chlorine production (as HCl) from PVC accelerates, hence the contact time between CAS constituents and HCl gas is quite short. Furthermore, the used PVC is in powder form, which means that the temperature is almost uniform across the particles leading undoubtedly to the simultaneous and instantaneous de-chlorination of the PVC.

As mentioned above, a neo-formed phase such as CuO produced from thermal decomposition [$Cu_2O(SeO_3)$] via Equation (5) occurred at apparent lower temperature (at 415 °C) attributed to the heat release from constituents' interaction in the sample mixture (CAS + PVC). Besides, the characteristic peaks of Cu_2O (d = 2.46 Å and d = 2.13 Å) appeared also in the XRD patterns of residues obtained at 415 °C and 505 °C exhibiting the effect of the carbon on the oxidation state of copper. These diffraction peaks are of low intensity for a temperature higher than 550 °C.

The XRD diffraction patterns of products for both CAS treatments are similar for the temperature higher than 505 °C (Tables 2 and 3). The XRD patterns of a CAS raw sample and its reaction product issued from the thermal treatment of CAS + PVC at 770 °C are compared in Figure 12. With a complex phase composition of a raw CAS sample ($CuAgSe$, $Cu_{2-x}Se_yS_{1-y}$, Ag_3AuSe_2 and SiO_2), the treatment residue at 770 °C is free of the selenium-bearing phase and it is composed of CuO, SiO_2 and Ag° (and Au°).

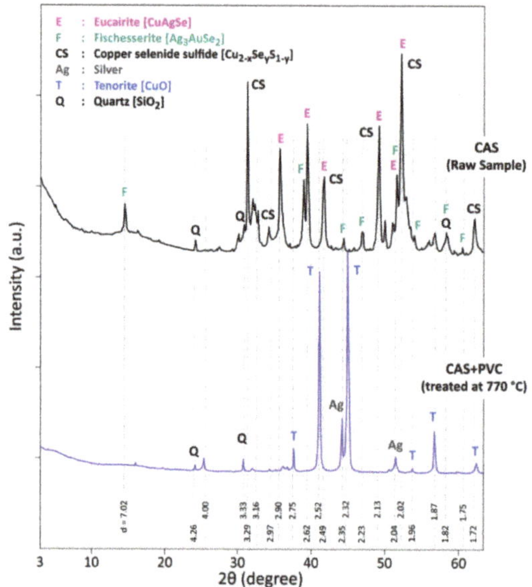

Figure 12. XRD diffractograms of a CAS raw sample and (CAS + PVC) thermal treatment product in air at 770 °C.

The EDS analysis of spot n° 1 (Figure 13) showed that Ag° (84.4 wt%) was the major constituent with some Au° (13.4 wt%) and Cu° (2.2 wt%). All smoothed particles had similar composition. The area noted by spot n° 2 was essentially composed of copper and oxygen with oxygen deficit to be CuO. The small area noted as n° 3 seemed to be composed of copper, silver and tellurium oxide (TeO_2).

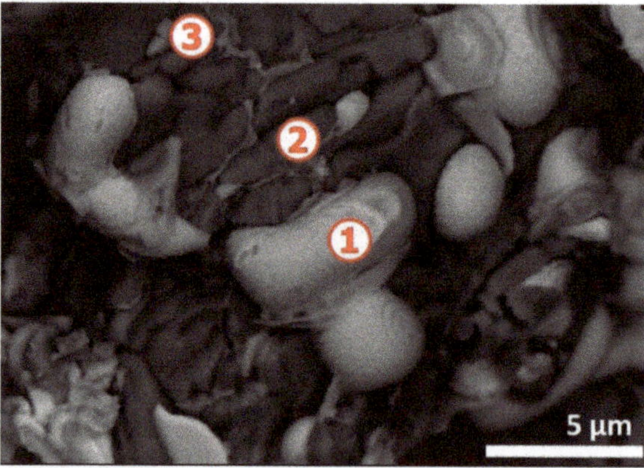

Figure 13. SEM aspects (BSE image) of the (CAS + PVC) sample treated at 770 °C in air atmosphere. Numbers 1 to 3 indicate the spots for microanalysis.

Likewise, the thermal treatment residues were examined by SEM-EDS. Figure 13 illustrates a typical morphology with a strong contrast indicating areas of distinct compositions. Data of SEM-EDX analysis of spots n° 1, 2 and 3 are reported in Table 4. The presence of well smoothed particles (image in Figure 13) indicates that the smelting of the metallic phase had been occurring during the treatment. Although the melting points of pure metals (Cu, Ag, Au) are higher than the temperature of the experimental tests, according to phase stability diagrams [40], the liquid phase appeared at temperatures lower than those of the fusion point of these metals. Furthermore, the heat release from the exothermic reactions can increase the sample temperature leading to local fusion of the treatment residue.

Table 4. Elemental composition (EDS data) of the (CAS + PVC) treated at 770°C in air atmosphere.

Elements	Spot n° 1		Spot n° 2		Spot n° 3	
	[1]wt%	[1]at%	wt%	at%	wt%	at%
O	-	-	12.85	37.29	4.38	17.85
Al	-	-	-	-	0.14	0.35
Si	-	-	-	-	0.11	0.26
Cu	2.16	3.84	83.92	61.32	60.52	62.04
Ag	84.40	88.45	3.23	1.39	18.40	11.11
Au	13.44	7.71	-	-	-	-
Te	-	-	-	-	16.45	8.40

[1] wt% and at% represents mass and atomic percentage, respectively.

The chosen copper anode slime sample contained a significant amount of selenium and tellurium, which belong to the scattered elements category [44]. As reported previously [45–47], these elements are used in thin films (CIGS—copper indium gallium selenide, and CdTe—cadmium telluride) and used in second-generation modules of photovoltaic panels. Recovery of these elements from by-products, end-of-life solar photovoltaics and other wasted materials will be a challenge of future research works to meet the volume of industrial demand.

These results encompassed some characteristics for the thermal treatment of a copper by-product in the presence of PVC and allowed us to understand the behavior of selected element compounds during the process. However, end-of-life materials (especially electric and electronic devices) are generally more complex in elemental, chemical and mineralogical composition of inorganics, with the presence of various organic matters and additives incorporated for enhancing the functional and commodity properties of the designed appliances.

The following section gives an idea about the interaction of bromine-bearing substances with valuable metals of an e-waste (printed circuit boards—PCBs).

3.4. Preliminary Results for Treatment of E-Waste in Air

Several selected research works [10,11,48–52] were devoted to the recovery of high value added metals and other components from e-waste by using various methods. A weak point for the metal recycling from e-waste is the high amounts of halogenated substances disturbing the thermal extraction process. Of particular interest are the data reported by Hino et al. [48] and Szałatkiewicz [51] with respect to the elemental and material compositions of the PCBs originating from discarded computers. The organic epoxy resin, inorganic glass fiber and metal elements represented 31.8 wt%, 37.6 wt% and 30.1 wt% of the PCBs, respectively. The bromine content in the organic substances was 5.07 wt%.

In the previous sections, it has been demonstrated that hydrogen chloride issued from PVC decomposition can react with several metal compounds generating their respective chlorides. Chlorine and bromine substances are often embedded in the plastic as flame retardants and they are still widely used despite environmental concerns. To gain an insight into the thermal behavior of the e-waste (PCBs of the obsolete end-of-life computers), isothermal tests under an air atmosphere were performed at 500 °C for reaction times of 0.25, 0.50, 0.75, 1.0 and 2.0 h. As the physical sample is composed of plate strips cut at about (10 × 2) cm, it was useful to repeat the experimental tests (3 times) at a given reaction time in order to have a reliable correlation between % ML of sample and the reaction time. This allowed us to overcome the impact of macroscopic heterogeneity with respect, at least, to plastic content. Additionally, for the reaction time of 1 h, 10 experimental tests of PBCs treatment in air atmosphere were performed to carefully check the reproducibility of experimental tests and to attain a better sampling of the obtained products. All the results are displayed in Figure 14.

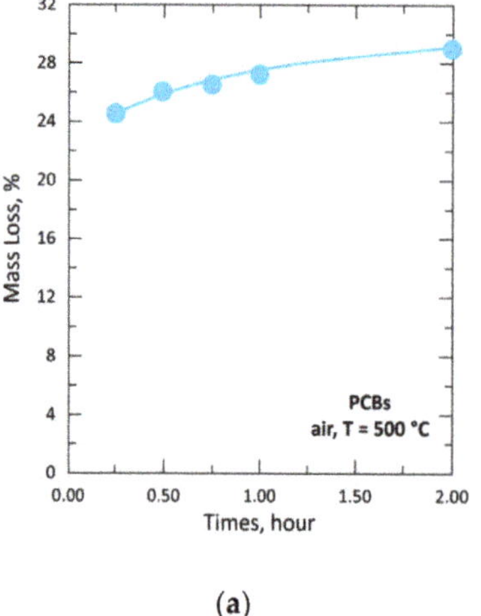 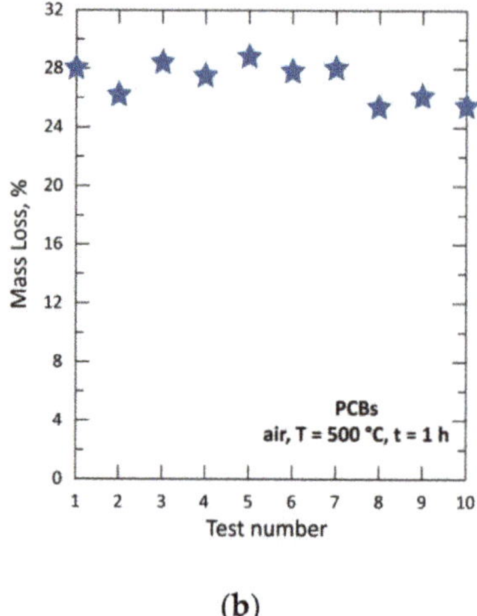

(a)   (b)

Figure 14. Mass change of the sample versus time for the treatment of printed circuit boards (PCBs) at 500 °C: (**a**) % mass loss (%ML) versus time; (**b**) %ML obtained for each test number at reaction time of 1 h.

About 24.5% ML were obtained during the treatment of PCBs sample for 0.25 h in air atmosphere (Figure 14a). The % ML of the sample increased slightly with reaction time to reach 29.0% during the treatment at 2.0 h. The experimental tests for 1 h gave mass losses oscillating between 25.5% and 29.0% with a mean value of 27.2% (Figure 14b). The small influence the treatment time had on the sample loss is probably due to the high reactivity of the organic matter towards air (O_2), therefore likely leading to a temperature increase in the reaction zone. The organic matter content (31.8%) of the PCBs from discarded computers previously reported [48,51] seems to be in good agreement with the results shown in Figure 14. The bromine content of the PCBs organic substances reported to be around 5 wt% in references [48,51] is assumed to apply to the similar e-waste used in the present study.

The treatment residue obtained for 1 h of treatment was sieved and the fine fraction (less than 210 µm) was analyzed by SEM-EDS technique. Likewise, the solid condensate was examined by the same technique.

A general SEM image of the treatment product is displayed in Figure 15a. Although most metals (especially Cu) are found in the coarse fraction of centimetric size, this fine fraction of the treatment product still contains Sn, Cu and Pb compounds (spectrum in Figure 15b). Si, Ca, Al and Mg are often found in the sticks (see SEM image of Figure 15a) being used as reinforced fibbers and/or glass weaves for the plastics.

Punctual analyses by EDS of various areas of SEM image of Figure 16 give different compositions as reported in Table 5. There are areas rich in lead (spot n° 1) and in tin (spot n° 2). Tantalum (spot n° 3) is almost certainly derived from tantalum capacitors. The EDS analyses of spot n° 4 represents mostly the elemental composition of an inorganic filler particle.

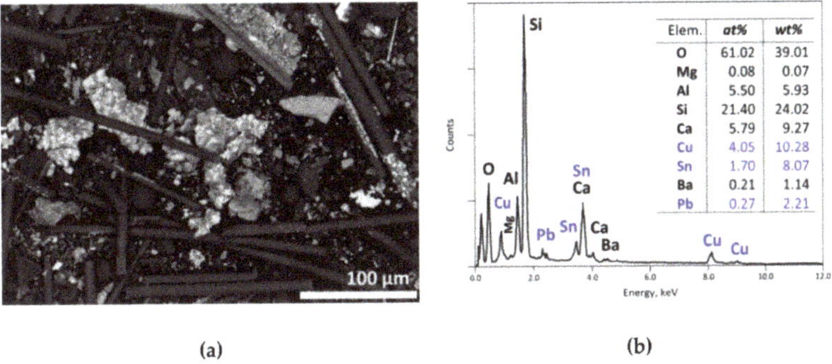

(a)   (b)

Figure 15. SEM-EDS results of a residue obtained from the treatment of the e-waste sample at 500 °C in air atmosphere: (**a**) general view (BSE micrograph) of the residue; (**b**) overall EDS analysis of the residue.

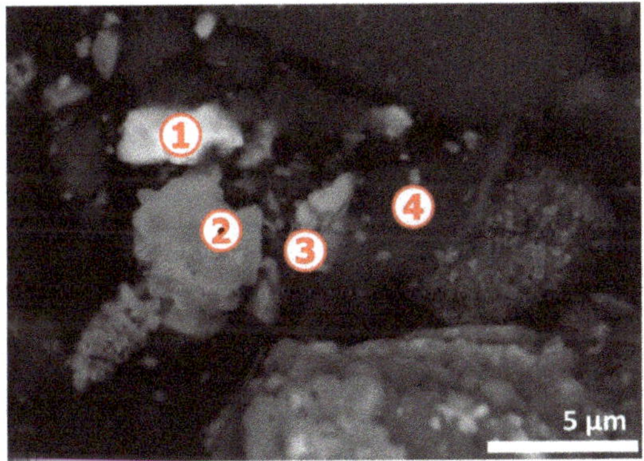

Figure 16. SEM aspects (BSE image) of a residue area obtained from the treatment of the e-waste sample at 500 °C in air atmosphere for 1 h. Numbers 1 to 4 indicate the spots for microanalysis.

Table 5. Elemental composition (EDS data) of the areas displayed in Figure 16.

Elements	Spot n° 1		Spot n° 2		Spot n° 3		Spot n° 4	
	[1]wt%	[1]at%	wt%	at%	wt%	at%	wt%	at%
O	1.89	15.62	21.39	65.63	20.75	66.29	46.96	66.27
Na	-	-	-	-	-	-	3.21	3.16
Al	-	-	0.65	1.19	1.64	3.11	11.92	9.97
Si	0.30	1.40	0.70	1.22	-	-	17.41	14.00
P	-	-	-	-	-	-	2.12	1.54
Ca	1.81	5.99	-	-	-	-	3.37	1.90
Ti	-	-	-	-	-	-	2.09	0.98
Mn	-	-	-	-	12.83	11.94	-	-
Cu	5.79	12.06	-	-	-	-	1.47	0.52
Br	7.13	11.82	-	-	-	-	-	-
Sn	-	-	77.26	31.96	2.38	1.02	5.02	0.95
Ta	-	-	-	-	62.40	17.63	-	-
Pb	83.08	53.11	-	-	-	-	6.43	0.70

[1] wt% and at% represent mass and atomic percentage, respectively.

The attractive result of the SEM-EDS analysis is given in Figure 17 and Table 6. Besides metals (Ag, Cu, Pb, Sn …), at various proportions, there is an appreciable amount of bromine (spots n° 1 to 3) indicating that bromine is bonded with part of the metals as bromides.

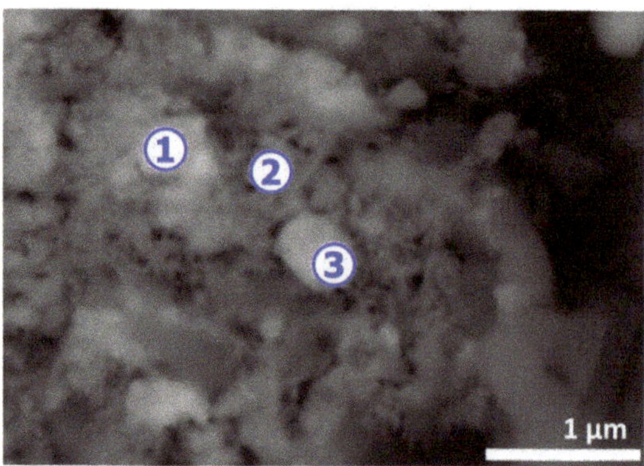

Figure 17. SEM aspects (BSE image) of a residue area obtained from the treatment of the e-waste sample at 500 °C in air atmosphere for 1 h. Numbers 1 to 3 indicate the spots for microanalysis.

Table 6. Elemental composition (EDS data) of the areas displayed in Figure 17.

Elements	Spot n° 1		Spot n° 2		Spot n° 3	
	[1]wt%	[1]at%	wt%	at%	wt%	at%
O	17.75	60.63	14.56	54.58	16.91	58.83
Cu	4.42	3.80	4.95	4.67	8.53	7.47
Br	9.88	6.75	16.52	12.39	16.84	11.73
Ag	4.39	2.22	9.73	5.41	2.57	1.32
Sn	49.28	22.69	33.07	16.71	28.60	13.41
W	4.18	1.24	2.99	0.98	2.92	0.88
Pb	10.11	2.67	18.17	5.26	23.64	6.35

[1] wt% and at% represents mass and atomic percentage, respectively.

As shown in Figure 18, gold, which is used in PCBs to assure circuit board contacts, seems to be inert to the treatment at this temperature.

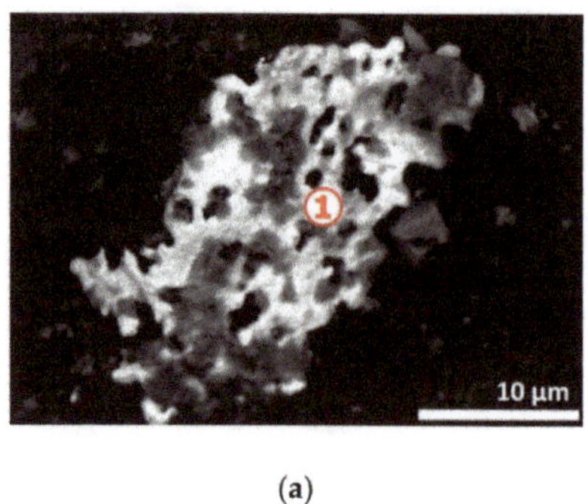
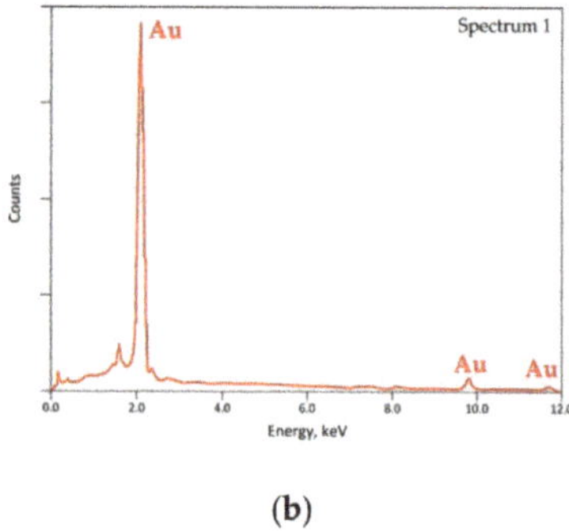

(**a**)

(**b**)

Figure 18. SEM-EDS results of a residue obtained from the treatment of the e-waste sample at 500 °C in air atmosphere: (**a**) general view (BSE micrograph) of a residue area; (**b**) punctual EDS spectrum.

A thermodynamic calculation [39] showed that the values of the $\Delta_r G°$ for the reaction (6), (7), (8) and (9) at 500 °C are −119.80, −124.11, −143.22 and −119.97 kJ/mol HBr, respectively. Such negative values suggest that the envisaged reactions are favorable from the thermodynamic point of view. The species AgBr, CuBr, PbBr$_2$ and SnBr$_4$ are chosen for their predominant thermodynamic stability in the studied system (Metal-O-Br) and based on the thermochemical data availability.

$$Ag_{(s)} + HBr_{(g)} + 1/4O_{2(g)} \rightarrow AgBr_{(l)} + 1/2H_2O_{(g)} \tag{6}$$

$$Cu_{(s)} + HBr_{(g)} + 1/4O_{2(g)} \rightarrow CuBr_{(l)} + 1/2H_2O_{(g)} \tag{7}$$

$$1/2Pb_{(s)} + HBr_{(g)} + 1/4O_{2(g)} \rightarrow 1/2PbBr_{2(l)} + 1/2H_2O_{(g)} \tag{8}$$

$$1/4Sn_{(s)} + HBr_{(g)} + 1/4O_{2(g)} \rightarrow 1/4SnBr_{4(g)} + 1/2H_2O_{(g)} \tag{9}$$

As indicated in Section 2, the exit gases were cooled resulting in the condensation of the vapor phase and the recovery of a solid condensate mixed often with unburnt carbonaceous matter, soot and liquid giving the visual appearance of a pasty mass. SEM-EDS results of a condensate are shown in Figure 19 with an EDS analysis of spot n° 1 corresponding roughly to PbBr$_2$. As the vapor pressure of PbBr$_2$ is near to 0.1 kPa [39], its presence in the condensate is probably due to it being brought back by the carrier gas and/or thanks to temperature increase enhancing its volatilization. Another EDS punctual analysis displayed in Figure 20b, noted as spot n° 2 of SEM image, suggested most likely the stannic bromide (SnBr$_4$) generated from the condensation of gaseous tin bromide [SnBr$_{4(g)}$] in the cooled part of the reactor. The obtained results showed that several metals could be concentrated in the vapor phase via volatilization of the neo-formed halides during the thermal process.

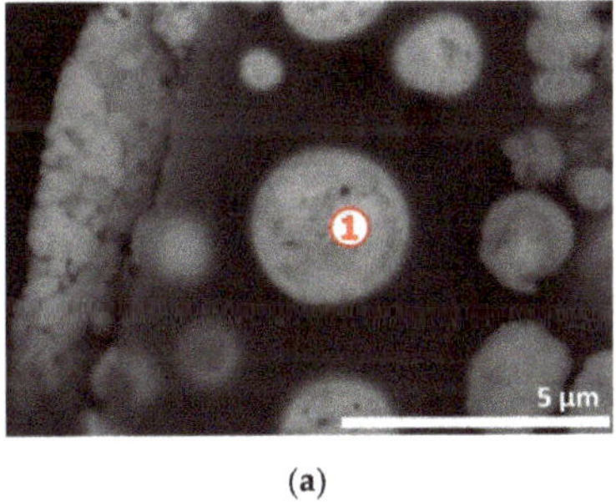

(a)

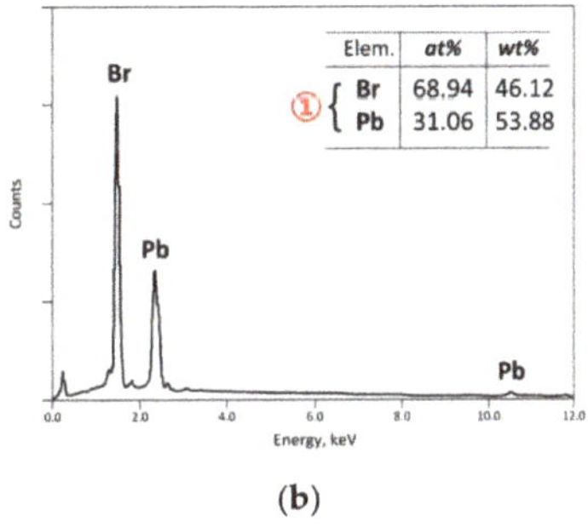

(b)

Figure 19. SEM-EDS results of a condensate issued from the e-waste sample treated at 500 °C in air atmosphere: (**a**) general view (BSE micrograph); (**b**) EDS analysis of spot n° 1.

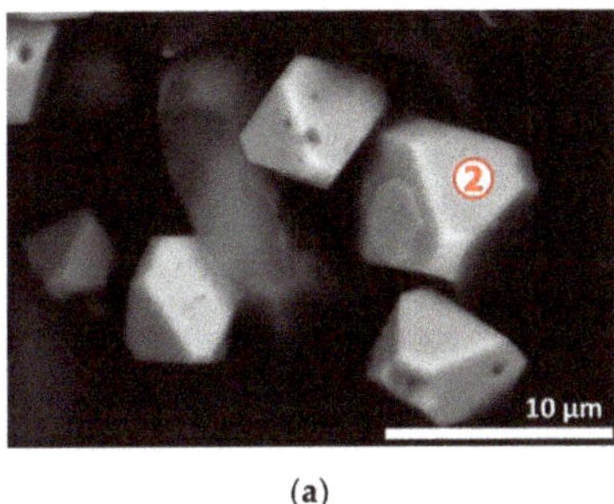

(a)

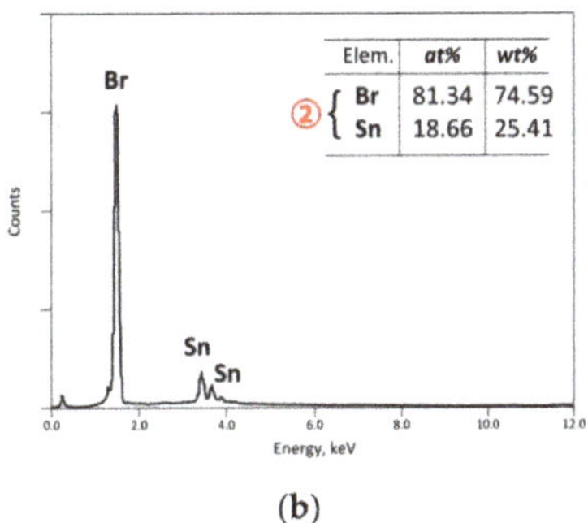

(b)

Figure 20. SEM-EDS results of a condensate issued from the e-waste sample treated at 500 °C in air atmosphere: (**a**) general view (BSE micrograph); (**b**) EDS analysis of spot n° 2.

As revealed during this study, the extraction of the critical and high-value elements from the chosen by-product and end-of-life materials represents a challenge due to the chemical and mineralogical complexity of their components as well as to the selectivity of the extractive chemistry of different metals for an efficient separation. Moreover, fair thermodynamic data, the ability to gather phases differentiation, punctual information about elemental content, in addition to monitoring the morphological and textural evolution of the thermally treated samples, are all contributing to a better understanding of the reaction mechanism involved and processing steps for the studied cases. Other aspects of this investigation, dealing with mass balance of the processes and assessment of technical and economical evaluation, should be carried out in future studies.

4. Conclusions

Copper Anode Slime (CAS) is a complex material containing the targeted elements (Ag, Au, Se, Te, Cu …) distributed in diverse mineralogical phases such as Ag_3AuSe_2, $CuAgSe$, $Cu_{2-x}Se_yS_{1-y}$.

The interaction of CAS constituents with HCl, issued from the degradation of polyvinyl chloride (PVC) under in an air atmosphere, started from about 200 °C leading to the favorable formation of AgCl and CuCl. Selenium is probably volatilized as selenium chlorides, most likely as $SeCl_{2(g)}$.

The isothermal degradation of PVC in an air atmosphere is a two-step process. The PVC de-chlorination occurs at temperature lower than 300 °C and proceeds with an apparent activation energy of around 120 kJ/mol. The second step should be attributed to the reaction of oxygen with carbonaceous matter generated in the first step. Full sample degradation and volatilization of the reaction products were achieved during treatment of PVC at about 500 °C for 1 h.

The reaction of HCl generated from PVC with CAS at temperature higher than 275 °C remained limited because the generation rate of HCl was higher than that of the interaction of HCl with CAS constituents.

The thermal treatment of CAS + PVC beyond 300 °C gave roughly similar results to the processing of CAS alone in air atmosphere. The formation of combined copper-selenium oxides, followed by their decomposition allowing solid copper oxides are the main steps of this treatment.

The final residue of the CAS + PVC treatment in air at temperature higher than 600 °C is composed of CuO (Cu_2O) and alloys of Ag-Au (with some Cu°), while tellurium is found as oxide (TeO_2), and silica (SiO_2) which is unreactive with respect to the thermal treatment. Further treatment of this product is required for the final separation of these high-value element compounds.

The bromine-bearing flame retardants of the printed circuits boards (PCBs) also appeared to be a good halogenating agent for several components of this e-waste. Bromides of copper, lead and tin, synthetized through the reaction of the evolved HBr with their respective metals, are distributed between residue and condensate of the PCBs treatment at 500 °C under air atmosphere. Further studies linked with various experimental parameters should be undertaken for a selective separation of the synthetized halides.

Author Contributions: Conceptualization, N.K., E.A., F.P. and J.Y; Formal analysis, S.S., S.D. and F.D.; Investigation, N.K., S.D. and F.D.; Visualization, S.S. and F.D.; Resources, N.K., F.P. and J.Y.; Writing—original draft, N.K., E.A., S.S. and F.P.; Writing—review and editing, N.K., E.A., F.P. and J.Y. All authors have read and agreed to the published version of the manuscript.

Funding: This research was carried out within the framework of French projects namely: *(i)*—PEPS-Urban mine-2014 (CNRS-INSIS); *(ii)*—"Investissements d'avenir"—ANR-10-LABX-21-01/LABEX RESSOURCES21; *(iii)*—ANR CARNOT programme.

Conflicts of Interest: The authors declare no conflict of interest.

References

1. Kanari, N.; Allain, E.; Shallari, S.; Diot, F.; Diliberto, S.; Patisson, F.; Yvon, J. Thermochemical route for extraction and recycling of critical, strategic and high value elements from by-products and end-of-life materials, Part I: Treatment of a copper by-product in air atmosphere. *Materials* **2019**, *12*, 1625. [CrossRef]
2. Critical raw materials—European Commission—Europa EU. Available online: http://ec.europa.eu/growth/sectors/raw-materials/specific-interest/critical_en (accessed on 15 March 2019).
3. Allain, E.; Kanari, N.; Diot, F.; Yvon, J. Development of a process for the concentration of the strategic tantalum and niobium oxides from tin slags. *Miner. Eng.* **2019**, *134*, 97–103. [CrossRef]
4. Holgersson, S.; Steenari, B.-M.; Björkman, M.; Cullbrand, K. Analysis of the metal content of small-size Waste Electric and Electronic Equipment (WEEE) printed circuit boards-part 1: Internet routers, mobile phones and smartphones. *Resour. Conserv. Recycl.* **2018**, *133*, 300–308. [CrossRef]
5. Kumar, S.; Rawat, S. Future e-waste: Standardisation for reliable assessment. *Gov. Inform. Q.* **2018**, *35*, S33–S42. [CrossRef]
6. Otto, S.; Kibbe, A.; Henn, L.; Hentschke, L.; Kaiser, F.G. The economy of e-waste collection at the individual level: A practice oriented approach of categorizing determinants of e-waste collection into behavioral costs and motivation. *J. Clean. Prod.* **2018**, *204*, 33–40. [CrossRef]
7. Jayaraman, K.; Vejayon, S.; Raman, S.; Mostafiz, I. The proposed e-waste management model from the conviction of individual laptop disposal practices-An empirical study in Malaysia. *J. Clean. Prod.* **2019**, *208*, 688–696. [CrossRef]
8. Abbondanza, M.N.M.; Souza, R.G. Estimating the generation of household e-waste in municipalities using primary data from surveys: A case study of Sao Jose dos Campos, Brazil. *Waste Manage.* **2019**, *85*, 374–384. [CrossRef] [PubMed]
9. Cho, B.-G.; Lee, J.-C.; Yoo, K. Valuable Metal Recycling. *Metals* **2018**, *8*, 345. [CrossRef]
10. Ding, Y.; Zhang, S.; Liu, B.; Zheng, H.; Chang, C.-C.; Ekberg, C. Recovery of precious metals from electronic waste and spent catalysts: A review. *Resour. Conserv. Recy.* **2019**, *141*, 284–298. [CrossRef]
11. Avarmaa, K.; Klemettinen, L.; O'Brien, H.; Taskinen, P. Urban mining of precious metals via oxidizing copper smelting. *Miner. Eng.* **2019**, *133*, 95–102. [CrossRef]
12. Amer, A.M. Processing of copper anodic-slimes for extraction of valuable metals. *Waste Manage.* **2003**, *23*, 763–770. [CrossRef]
13. Khaleghi, A.; Ghader, S.; Afzali, D. Ag recovery from copper anode slime by acid leaching at atmospheric pressure to synthesize silver nanoparticles. *Int. J. Min. Sci. Technol.* **2014**, *24*, 251–257. [CrossRef]
14. Li, X.J.; Yang, H.Y.; Jin, Z.N.; Chen, G.B.; Tong, L.L. Transformation of selenium-containing phases in copper anode slimes during leaching. *JOM* **2017**, *69*, 1932–1938. [CrossRef]
15. Xiao, L.; Wang, Y.L.; Yu, Y.; Fu, G.Y.; Han, P.W.; Sun, Z.H.I.; Ye, S.F. An environmentally friendly process to selectively recover silver from copper anode slime. *J. Clean. Prod.* **2018**, *187*, 708–716. [CrossRef]
16. Kilic, Y.; Kartal, G.; Timur, S. An investigation of copper and selenium recovery from copper anode slimes. *Int. J. Miner. Process.* **2013**, *124*, 75–82. [CrossRef]
17. Wang, Z.; Wei, R.; Wang, X.; He, J.; Wang, J. Pyrolysis and combustion of polyvinyl chloride (PVC) sheath for new and aged cables via thermogravimetric analysis-Fourier transform infrared (TG-FTIR) and calorimeter. *Materials* **2018**, *11*, 1997. [CrossRef]
18. Niu, L.; Xu, J.; Yang, W.; Ma, J.; Zhao, J.; Kang, C.; Su, J. Study on the synergetic fire-retardant effect of nano-Sb_2O_3 in PBT Matrix. *Materials* **2018**, *11*, 1060. [CrossRef]
19. Rani, M.; Marchesi, C.; Federici, S.; Rovelli, G.; Alessandri, I.; Vassalini, I.; Ducoli, S.; Borgese, L.; Zacco, A.; Bilo, F.; et al. Miniaturized near-infrared (MicroNIR) spectrometer in plastic waste sorting. *Materials* **2019**, *12*, 2740. [CrossRef]
20. Hermosillo-Nevárez, J.J.; Bustos-Terrones, V.; Bustos-Terrones, Y.A.; Uriarte-Aceves, P.M.; Rangel-Peraza, J.G. Feasibility study on the use of recycled polymers for malathion adsorption: Isotherms and kinetic modeling. *Materials* **2020**, *13*, 1824. [CrossRef]
21. Mun, S.-Y.; Hwang, C.-H. Experimental and numerical studies on major pyrolysis properties of flame retardant PVC cables composed of multiple materials. *Materials* **2020**, *13*, 1712. [CrossRef]
22. Kaczorek-Chrobak, K.; Fangrat, J. PVC-based copper electric wires under various fire conditions: Toxicity of fire effluents. *Materials* **2020**, *13*, 1111. [CrossRef] [PubMed]

23. Allain, E.; Gaballah, I.; Djona, M. Extraction of tantalum and niobium from tin slags by chlorination and carbochlorination. *Metall. Mat. Trans. B* **1997**, *28B*, 359–369. [CrossRef]

24. Kanari, N.; Allain, E.; Gaballah, I. Reactions of wüstite and hematite with different chlorinating agents. *Thermochim. Acta* **1999**, *335*, 79–86. [CrossRef]

25. Kanari, N.; Gaballah, I.; Allain, E. Kinetics of oxychlorination of chromite Part I. Effect of temperature. *Thermochim. Acta* **2001**, *371*, 143–154. [CrossRef]

26. Kanari, N.; Gaballah, I.; Allain, E. Kinetics of oxychlorination of chromite part II. Effect of reactive gases. *Thermochim. Acta* **2001**, *371*, 75–86. [CrossRef]

27. Kanari, N.; Allain, E.; Joussemet, R.; Mochón, J.; Ruiz-Bustinza, I.; Gaballah, I. An overview study of chlorination reactions applied to the primary extraction and recycling of metals and to the synthesis of new reagents. *Thermochim. Acta* **2009**, *495*, 42–50. [CrossRef]

28. Kanari, N.; Menad, N.; Diot, F.; Allain, E.; Yvon, J. Phosphate valorization by dry chlorination route. *J. Min. Metall. Sect. B Metall.* **2016**, *52*, 17–24. [CrossRef]

29. Kanari, N.; Menad, N.-E.; Ostrosi, E.; Shallari, S.; Diot, F.; Allain, E.; Yvon, J. Thermal behavior of hydrated iron sulfate in various atmospheres. *Metals* **2018**, *8*, 1084. [CrossRef]

30. Kanari, N.; Evrard, O.; Neveux, N.; Ninane, L. Recycling ferrous sulfate via super-oxidant synthesis. *JOM* **2001**, *53*, 32–33. [CrossRef]

31. Kanari, N.; Ostrosi, E.; Ninane, L.; Neveux, N.; Evrard, O. Synthesizing alkali ferrates using a waste as a raw material. *JOM* **2005**, *57*, 39–42. [CrossRef]

32. Kanari, N.; Filippov, L.; Diot, F.; Mochón, J.; Ruiz-Bustinza, I.; Allain, E.; Yvon, J. Synthesis of potassium ferrate using residual ferrous sulfate as iron bearing material. *J. Phys. Conf. Ser.* **2013**, *416*, art. no. 012013. [CrossRef]

33. Kanari, N.; Filippova, I.; Diot, F.; Mochón, J.; Ruiz-Bustinza, I.; Allain, E.; Yvon, J. Utilization of a waste from titanium oxide industry for the synthesis of sodium ferrate by gas-solid reactions. *Thermochim. Acta* **2014**, *575*, 219–225. [CrossRef]

34. Kanari, N.; Ostrosi, E.; Diliberto, C.; Filippova, I.; Shallari, S.; Allain, E.; Diot, F.; Patisson, F.; Yvon, J. Green process for industrial waste transformation into super-oxidizing materials named alkali metal ferrates (VI). *Materials* **2019**, *12*, 1977. [CrossRef] [PubMed]

35. Chen, T.T.; Dutrizac, J.E. A mineralogical study of the deportment and reaction of silver during copper electrorefining. *Metall. Mat. Trans. B* **1989**, *20B*, 345–361. [CrossRef]

36. Petkova, E.N. Microscopic examination of copper electrorefining slimes. *Hydrometallurgy* **1990**, *24*, 351–359. [CrossRef]

37. Petkova, E.N. Hypothesis about the origin of copper electrorefining slime. *Hydrometallurgy* **1994**, *34*, 343–358. [CrossRef]

38. Chen, T.T.; Dutrizac, J.E. Mineralogical characterization of a copper anode and the anode slimes from the La Caridad copper refinery of Mexicana de Cobre. *Metall. Mat. Trans. B* **2005**, *36B*, 229–240. [CrossRef]

39. Roine, A. *Outokumpu HSC Chemistry for Windows, Version 3.0*; Outokumpu Research: Pori, Finland, 1997.

40. ASM Handbook—Volume 3, Alloy Phase Diagrams. Available online: http://s1.iran-mavad.com/ASM%20hanbooks/Vol_3_ASM%20handbooks_iran-mavad.com.pdf (accessed on 7 July 2020).

41. Kanari, N.; Gaballah, I.; Allain, E.; Menad, N. Chlorination of chalcopyrite concentrates. *Metall. Mat. Trans. B* **1999**, *30*, 567–576. [CrossRef]

42. Kanari, N.; Gaballah, I.; Allain, E. A low temperature chlorination–volatilization process for the treatment of chalcopyrite concentrates. *Thermochim. Acta* **2001**, *373*, 75–93. [CrossRef]

43. Fokina, E.L.; Klimova, E.V.; Charykova, M.V.; Krivovichev, V.G.; Platonova, N.V.; Semenova, V.V.; Depmeier, W. The thermodynamics of arsenates, selenites, and sulfates in the oxidation zone of sulfide ores: VIII. Field of thermal stability of synthetic analog of chalcomenite, its dehydration and dissociation. *Geol. Ore Deposit.* **2014**, *56*, 538–545. [CrossRef]

44. Zhang, L.; Xu, Z. A critical review of material flow, recycling technologies, challenges and future strategy for scattered metals from minerals to wastes. *J. Clean. Prod.* **2018**, *202*, 1001–1025. [CrossRef]

45. Davidsson, S.; Höök, M. Material requirements and availability for multi-terawatt deployment of photovoltaics. *Energy Policy* **2017**, *108*, 574–582. [CrossRef]

46. Padoan, F.C.S.M.; Altimari, P.; Pagnanelli, F. Recycling of end of life photovoltaic panels: A chemical prospective on process development. *Sol. Energy* **2019**, *177*, 746–761. [CrossRef]

47. Domínguez, A.; Geyer, R. Photovoltaic waste assessment of major photovoltaic installations in the United States of America. *Renew. Energy* **2019**, *133*, 1188–1200. [CrossRef]
48. Hino, T.; Agawa, R.; Moriya, Y.; Nishida, M.; Tsugita, Y.; Araki, T. Techniques to separate metal from waste printed circuit boards from discarded personal computers. *J. Matter Cycles Waste Manag.* **2009**, *11*, 42–54. [CrossRef]
49. Bizzo, W.A.; Figueiredo, R.A.; De Andrade, V.F. Characterization of printed circuit boards for metal and energy recovery after milling and mechanical separation. *Materials* **2014**, *7*, 4555–4566. [CrossRef]
50. Chiang, H.-L.; Lin, K.-H. Exhaust constituent emission factors of printed circuit board pyrolysis processes and its exhaust control. *J. Hazard. Mater.* **2014**, *264*, 545–551. [CrossRef]
51. Szałatkiewicz, J. Metals recovery from artificial ore in case of printed circuit boards, using plasmatron plasma reactor. *Materials* **2016**, *9*, 683. [CrossRef]
52. Wan, X.; Fellman, J.; Jokilaakso, A.; Klemettinen, L.; Marjakoski, M. Behavior of waste printed circuit board (WPCB) materials in the copper matte smelting process. *Metals* **2018**, *8*, 887. [CrossRef]

© 2020 by the authors. Licensee MDPI, Basel, Switzerland. This article is an open access article distributed under the terms and conditions of the Creative Commons Attribution (CC BY) license (http://creativecommons.org/licenses/by/4.0/).

Article

Green Process for Industrial Waste Transformation into Super-Oxidizing Materials Named Alkali Metal Ferrates (VI)

Ndue Kanari [1,*], Etleva Ostrosi [2], Cécile Diliberto [3], Inna Filippova [1], Seit Shallari [4], Eric Allain [1], Frederic Diot [1], Fabrice Patisson [5] and Jacques Yvon [1]

[1] GeoRessources Laboratory, UMR 7359 CNRS, CREGU, Université de Lorraine, 2, rue du doyen Roubault, BP 10162, 54505 Vandoeuvre-lès-Nancy, France; inna.filippova@univ-lorraine.fr (I.F.); ericgallain@gmail.com (E.A.); frederic.diot@univ-lorraine.fr (F.D.); jacques.yvon@univ-lorraine.fr (J.Y.)

[2] Ville de Montréal, Direction de l'environnement, Division de la Planification et du Suivi Environnemental, 801, rue Brennan, Montréal, QC H3C 0G4, Canada; etlevao@yahoo.com

[3] Institut Jean Lamour, UMR 7198 CNRS, Université de Lorraine, Equipe 'Matériaux pour le Génie Civil', IUTNB, BP 90137, 54600 Villers-lès-Nancy, France; cecile.diliberto@univ-lorraine.fr

[4] Agricultural University of Tirana, Faculty of Agriculture and Environment, 1029 Tirana, Albania; seitshallari@gmail.com

[5] Institut Jean Lamour, UMR 7198 CNRS, Labex DAMAS, Université de Lorraine, Campus Artem, 2 allée André Guinier, BP 50840, 54011 Nancy, France; fabrice.patisson@univ-lorraine.fr

* Correspondence: ndue.kanari@univ-lorraine.fr; Tel.: +33-372-744-530

Received: 29 May 2019; Accepted: 17 June 2019; Published: 19 June 2019

Abstract: The investigation presented here features the design of a cleaner and greener chemical process for the conversion of industrial wastes into super-oxidizing materials. The waste of interest is the iron sulfate heptahydrate ($FeSO_4 \cdot 7H_2O$) mainly generated through the sulfate route used for titanium dioxide industrial production. The products of this transformation process are alkali ferrates (A_2FeO_4, A = Na, K) containing iron in its hexavalent state and considered as powerful oxidants characterized by properties useful for cleaning waters, wastewaters, and industrial effluents. The proposed process includes two steps: (i) The first step consisting of the pre-mixing of two solids (AOH with $FeSO_4 \cdot xH_2O$) in a rotary reactor allowing the coating of iron sulfate in the alkali hydroxides through solid–solid reactions; and (ii) the second step involves the synthesis of alkali ferrates in a fluidized bed by oxidation of the single solid obtained in the first step in diluted chlorine. The chemical synthesis of alkali ferrates can be carried out within a timeframe of a few minutes. The usage of a fluidized bed enhanced the energy and mass transfer allowing a quasi-complete control of the ferrate synthesis process. The alkali ferrate synthesis process described here possesses many characteristics aligned with the principles of the "green chemistry".

Keywords: industrial waste; alkali ferrates; super-oxidizing materials; fluidized bed; green process

---

1. Introduction

Iron compounds are abundant in most nonferrous metal deposits where these metals often represent a small fraction. Although part of the ferrous components are separated during mining and primary mineral processing, a large part of iron goes through the extractive processes of these metals. Further, the iron is sometimes present in the natural mineral bodies (e.g., $FeTiO_3$, $CuFeS_2$, $(Ni,Fe)_9S_8$) of target metals that can be separated during the primary metal extraction. Therefore, considerable amounts of iron bearing co-products and wastes are inevitably generated from both hydro- and pyro-metallurgical operations on such raw materials.

One typical industrial example of generation of iron-containing waste is the extraction of titanium oxide (TiO_2) from its bearing materials such as ilmenite, rutile, anatase, and slags. As described previously [1], "chloride" and "sulfate" routes are processes currently used for TiO_2 production. The sulfate process, using ilmenite ($FeTiO_3$) as a raw material, can lead to the generation of amounts of iron sulfate ($FeSO_4 \cdot 7H_2O$—melanterite) as high as 6 tons of $FeSO_4 \cdot 7H_2O$ per ton of produced TiO_2 [2], which is a real environmental drawback for the TiO_2 industry. By 1993, all countries producing TiO_2 through the sulfate process had to comply with the European directive [3] in order to avoid dumping industrial waste (such as melanterite) into the seawater. In the perspective of a circular economy and sustainable development, the ideal scenario should be to consider these wastes (e.g., iron sulfate) as an input for new chemicals and material synthesis, which are intrinsically non-hazardous, thus excluding the notion of undesirable by-product. Few recent investigations [4–7] address the usage of melanterite in various application and its treatment, such as for synthesizing slow-release fertilizers [5], cation-substituted $LiFePO_4$ [6], and for Fe_3O_4 production through reductive decomposition using pyrite [7].

As a continuation of our previous research investigations [2,8–10], the present work aims at transforming industrial iron sulfate into alkali metal ferrates (A_2FeO_4, A = Na, K) considered as useful materials in different fields. One may note that the denomination "ferrate" is generally attributed to compounds containing iron at an oxidation state higher than Fe(III) and the ferrates(VI) seem to be the best known and most studied [2]. In an aqueous solution, the ferrate ion (FeO_4^{2-}) is reduced, generating both $Fe(OH)_3$ and nascent oxygen (Equation (1)). As summarized early [2], ferrates are used for water treatment due to their powerful oxidizing capacity ($E(FeO_4^{2-}/Fe^{3+}$ = 2.2 V)) (oxidation of organic and mineral materials, bactericide agent) and because of the flocculating property of the evolved $Fe(OH)_3$. The ferrates can replace chlorine in the pre-oxidation stage of water and can partially be used as a substitute of the iron and aluminum salts used as coagulating and flocculating agents. Furthermore, the decomposition of alkali ferrates generates basic medium favorable for the precipitation of heavy metals. These properties existing together in one single compound make ferrates an ongoing material particularly interesting for water treatment and effluent cleaning. Moreover, the oxidation products (iron oxy/hydroxides) (Equation (1)) are inoffensive to the environment.

$$2FeO_4^{2-} + 5H_2O \rightarrow 2Fe(OH)_3 + 4OH^- + 3O^\bullet \tag{1}$$

The research works performed by Fremy [11–13] are frequently cited in the literature as the first ones to scientifically reveal the existence of iron in a hexavalent state and to effectively achieve ferrates synthesis. Since then, the preparation methods, developed mostly on a laboratory scale, have made little progress, and can be classified into three groups:

- The high temperature route consisting of heating and/or melting various iron oxides bearing materials under high concentration of alkali substances and oxygen flow [14–16]. These synthesis methods, performed at temperatures as high as 800 °C, seem to be mostly ineffective since the Fe(VI) is not stable at temperatures higher than 200 °C. Most probably, the Fe(VI) resulted from the dismutation of synthetized Fe(IV) and/or Fe(V) during the manipulation with the synthesis product.

- The wet/humid oxidation of Fe(III) salts bearing solutions, under strong alkaline conditions, using hypochlorite or chlorine as oxidant. This method is the most used since the 1950s [17–20]. One of the drawbacks is that the wet method used pure chemicals and required many operations for Fe(VI) preparation and separation, making it very costly. Moreover, the water reacts with ferrate (Equation (1)) leading to its reduction into Fe(III).

- The electrochemical method [21–23], allowing the oxidation of anode (iron or its alloys) by operating mainly in concentrated NaOH and/or KOH solutions. However, the decomposition of Fe(VI) by water, low current efficiency, and anode passivation are some of the concerns for this route.

Synthesis of alkali ferrates (AF) by gas–solid reactions performed in a rotary reactor using chlorine as an oxidant showed that synthesis was achieved without external heat supply [2,8–10]. However, it was observed that the synthesis reactions were highly exothermic leading to a temperature rise in the reaction zone above 150 °C, when only 10 g of solids were used for the potassium and/or sodium ferrate synthesis. Experimental results indicated that the heat generation provoked the sample agglomeration and dramatically decreased the Fe(VI) efficiency of the synthesis process. The temperature increase phenomena became a real "obstacle" when higher amounts (100 g) of solids were used for the AF synthesis, although the reactor was cooled. Further, the kinetic of Na-ferrate synthesis is expected to be low compared with that of potassium ferrate [2]. In other words, it was concluded that the synthesis of alkali ferrates in a rotary reactor would not be considered as an appropriate route for an eventual AF large-scale production.

In this context, the goal of this research work is threefold: (i) Use the industrial iron sulfate as raw material for the AF synthesis; (ii) develop an appropriate process for AF preparation at room temperature; (iii) optimize the process by data analysis of various parameters affecting the efficiency of Fe(II,III) conversion into Fe(VI). The AF synthesis (mostly sodium ferrate preparation) is achieved in fluidized bed (FB) which can be considered as an appropriate reactor for handling processes that require high energy and mass transfers. It should be noted that the alkali metal ferrates (VI) manufacturing process, as developed in this work, is unique in its field.

2. Materials and Methods

Industrial ferrous sulfate (mainly monohydrate) is used for the synthesis of alkali ferrates. An examination by scanning electron microscope (Hitachi Ltd., Tokyo, Japan) coupled with energy dispersive spectrometry (Kevex Corp., Foster, CA, USA) (SEM-EDS), X-ray diffraction (XRD), chemical analyses, and Mössbauer spectroscopy (MS) suggested that the sample is free of heavy metals and that the quasi-totality of iron is in a bivalent state as $FeSO_4 \cdot H_2O$. Another sample of ferrous sulfate heptahydrate is also used after dehydration in an oven at about 150 °C leading to the formation of $FeSO_4 \cdot H_2O$ and $FeSO_4 \cdot OH$ as the main iron-bearing phases. Commercial sodium hydroxide is used as pellets of 2 mm and pearls of about 1 mm. The oxidizing agent (chlorine) and diluting gas (nitrogen) were of high purity, whilst the air was supplied by a compressor.

The flowchart summarizing the features of the experimental protocol established for the synthesis of alkali ferrates is schematized in Figure 1. Accordingly, the synthesis process consisted of two main steps. The first step consists of a premixing of mostly $FeSO_4 \cdot H_2O$ and/or $FeSO_4 \cdot OH$ with NaOH resulting in a single solid. This step is realized by using the experimental setup represented in Figure 2. The parameters studied for the solid premixing step are related to physical characteristics of iron sulfate and NaOH feed, their amounts, as well as the premixing time.

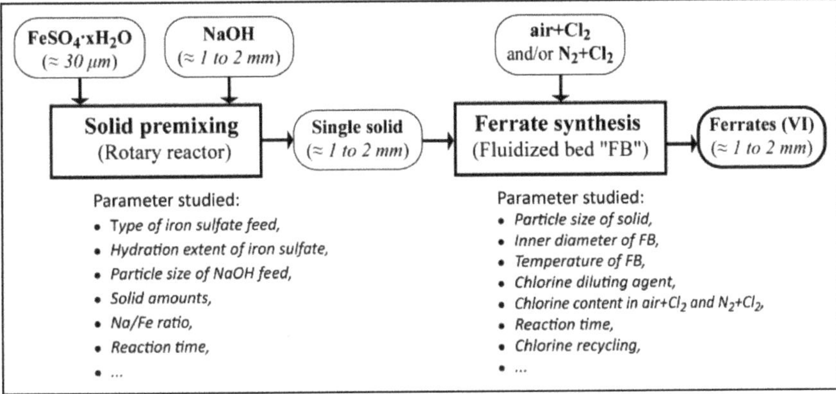

Figure 1. Schematic representation of the experimental procedure applied for the ferrates(VI) synthesis through a two-step process.

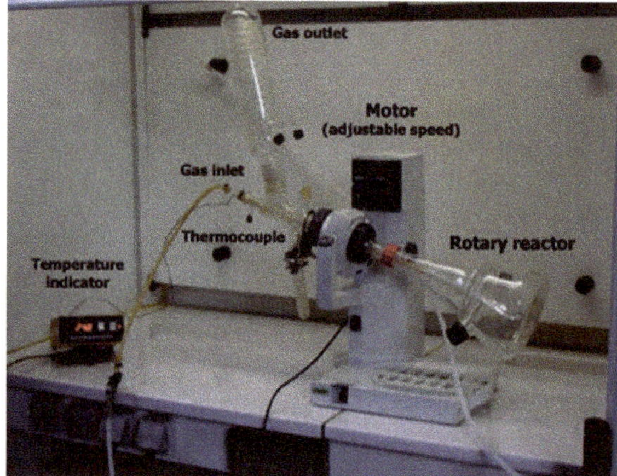

Figure 2. Apparatus assembly with a baffled reactor used for mixing iron sulfate with sodium hydroxide.

The second step of the synthesis process (Figure 1) consists of the reaction of the granulated and single solid, prepared in step 1, with chlorine diluted in air and/or N_2 leading to the sodium ferrate preparation. A fluidized bed reactor equipped with a temperature-regulated water system is used for the experimental tests of alkali ferrate synthesis. A second fluidized bed is added in series for recycling the unreacted oxidant (chlorine) during AF synthesis taking place in the first fluidized bed. Several experimental parameters related to the ferrate synthesis step (particle size of solid, temperature, chlorine content, reaction time) were investigated. Other details of the experimental procedure will be introduced when describing the experimental results for both steps.

Solid synthesis products are subjected to visible microscopy, SEM-EDS, and XRD to examine the structure and the composition of the solid reaction products. The procedure of these characterization techniques is given in a recently published material [24].

Mössbauer spectroscopy was used to seek information about the oxidation state of iron as well as to evaluate the Fe(VI) content in the synthesis product. Details about this analysis method were given earlier by Jeannot et al. [25]. However, this method of examination is time consuming, i.e., the acquisition of a Mössbauer spectrum may take up to 24 h. In this context, a chemical analysis method is performed to quickly determine the Fe(VI) synthesis efficiency of the ferrate synthesis trials. This method is based on the chemical reaction of Fe(VI) with an excess of ferrous sulfate solution; then, the excess of Fe(II) is titrated with potassium bichromate.

3. Results and Discussion

3.1. Concept of the Vibrating Fluidized Bed

The fluidized bed technique is very attractive for different processes related to gas–solid reactions and for its easy extrapolation. However, a fluidized bed is not suitable for solids with important differences in particles sizes (or density), as it is the case of NaOH and iron sulfate. As a reminder, the mean particle sizes of NaOH, found in market, were about 1, 2, and 5 mm and that of $FeSO_4 \cdot H_2O$ is less than 100 µm. A conventional fluidization of NaOH and $FeSO_4 \cdot H_2O$ (for a given fluidization velocity) allows a heterogeneous distribution of the solids in the reactor. It was suggested that the vibrating fluidized bed would be a solution for the homogeneous fluidization of NaOH and iron sulfate. The distribution of particles in the bed through the respective conditions of the fluidization alone, vibration alone, and fluidization coupled with vibration are schematized in Figure 3. The choice of the optimal fluidization velocity of the reactive gases for the iron sulfate creates the distribution situation described in Figure 3a. Thus, iron sulfate is fluidized when the NaOH pearls creates a packed bed at the bottom of the reactor. If the bed is only vibrated (Figure 3b), the pearls of NaOH move towards the top, while the iron sulfate accumulates at the bottom of the reactor. Meanwhile, the use of blow and vibration leads to an almost homogenous 'fluidization' of both solids (FeSO$_4$ and NaOH) as described in Figure 3c. However, it seems that the vibrating fluidized bed is difficult to control on an industrial scale production, especially for delicate processes such as ferrate synthesis.

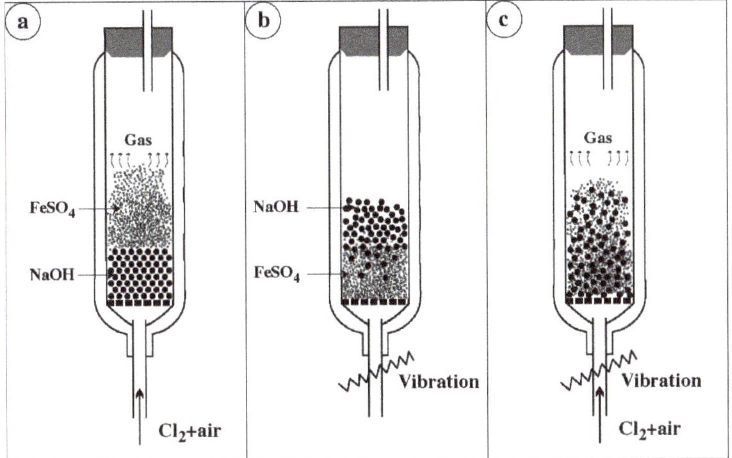

Figure 3. Schematic particles distribution of solids in a vibrated fluidized bed: (**a**) optimal gas fluidization velocity for iron sulfate; (**b**) only vibration of fluidized bed; (**c**) simultaneous blowing and vibration of fluidized bed.

3.2. Idea of Premixing of Solids Prior to Fluidization

The overall reaction of the sodium ferrate synthesis can be described by Equation (2), although the exact formula of this compound seems to be still undefined. It was already confirmed experimentally that the synthesis of Na-ferrate via Equation (2) is exothermic. The possible "two by two" reactions of the three substances (FeSO$_4$·H$_2$O, Cl$_2$, and NaOH) could be represented by Equations (3)–(5). Iron sulfate monohydrate does not react with chlorine at room temperature. Recent experience in the field of gaseous chlorine reactivity with iron compounds showed that the oxidation of Fe(II) of wüstite (FeO) into Fe(III) takes place at temperatures higher than 200 °C [26]. Chlorine in the presence of oxygen can oxidize Cr(III) into Cr(VI), generating chromium oxychloride [27–29].

As could be expected, the reaction of chlorine with NaOH produces NaCl as a final reaction product involving heat with $\Delta H = -128$ kJ/mol NaOH [30]. The most interesting reaction to be considered is that of iron sulfate with sodium hydroxide (Equation (5)) resulting in the formation of Na$_2$SO$_4$. The sodium sulfate is also an unavoidable product of the Na-ferrate synthesis (Equation (2)). Consequently, it was suggested to react iron sulfate with NaOH, allowing them to form a single solid (mixture of Fe(OH)$_2$ and Na$_2$SO$_4$), which would be suitable for the subsequent fluidization in the FB. The following paragraphs will describe the experimental results of the solid premixing concept.

$$\text{FeSO}_4\cdot\text{H}_2\text{O} + 2\text{Cl}_2 + 8\text{NaOH} \rightarrow \text{Na}_2\text{FeO}_4 + \text{Na}_2\text{SO}_4 + 4\text{NaCl} + 5\text{H}_2\text{O} \quad (2)$$

$$\text{FeSO}_4\cdot\text{H}_2\text{O} + \text{Cl}_2 \rightarrow \text{No evident reaction at room temperature} \quad (3)$$

$$2\text{NaOH} + \text{Cl}_2 \rightarrow 2\text{NaCl} + \text{H}_2\text{O} + 0.5\text{O}_2 \quad (4)$$

$$\text{FeSO}_4\cdot\text{H}_2\text{O} + 2\text{NaOH} \rightarrow \text{Fe(OH)}_2 + \text{Na}_2\text{SO}_4 + \text{H}_2\text{O} \quad (5)$$

3.3. Premixing of NaOH Pellets (2 mm) with Iron Sulfate

Sodium hydroxide conditioned as 2-mm pellets was used for the experimental tests. Iron sulfate monohydrate was chosen as iron salt for the same tests. The reaction of these solids was carried out in a rotary reactor of 2.5 L without presence of any gas and using an apparatus of premixing step as illustrated in Figure 2. About 170 g of solids with a molar ratio Na/Fe close to 8 (to satisfy Equation (2)) were loaded in the reactor and rotated at a speed of 20 rpm. The premixing time was fixed at 30 min. The evolution of temperature during the reaction was recorded and the data are displayed as plots of temperature versus time in Figure 4.

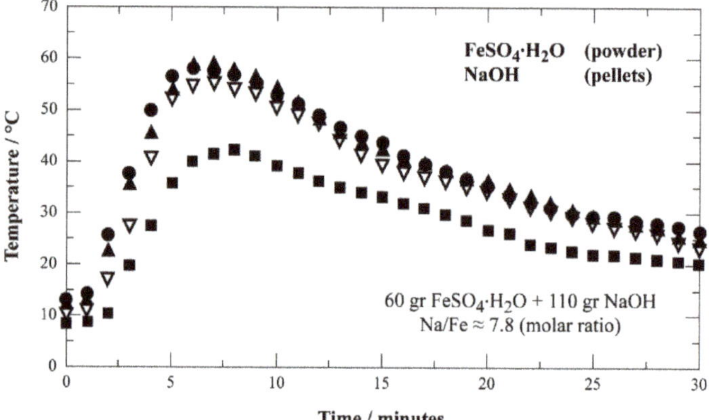

Figure 4. Evolution of the temperature as a function of time for the reaction of FeSO$_4$·H$_2$O during their mixing in the rotary reactor (four trials run in same conditions).

As shown by this figure, the reaction is exothermic leading to a temperature of about 60 °C in the reaction zone. The NaOH pellets become gray-black, but they more or less keep their initial shapes. Images of initial substances NaOH pellets (visible microscopy (VM)) and FeSO$_4$·H$_2$O (SEM) are shown in Figure 5a,b, respectively. Figure 5a clearly shows that NaOH consists of spherical particles of a diameter lower than or equal to 2 mm. As a contrast, FeSO$_4$·H$_2$O is composed of grains of different shapes with an equivalent diameter lower than 30 μm (Figure 5b).

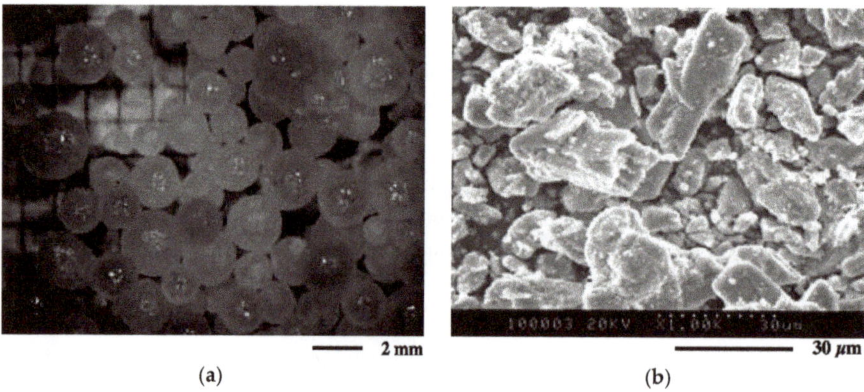

Figure 5. View of sodium hydroxide and ferrous sulfate samples: (**a**) Visible microscopy (VM) images of NaOH pellets; (**b**) SEM images of FeSO$_4$·H$_2$O.

Some relevant information about the reaction of NaOH with $FeSO_4 \cdot H_2O$ was revealed by SEM-EDS investigation. A representative NaOH grain after reaction with iron sulfate (image SEM) and its elemental analyses (EDS) are grouped in Figure 6.

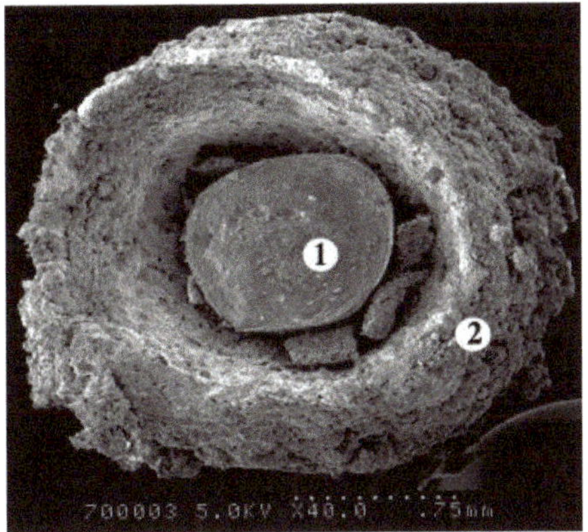

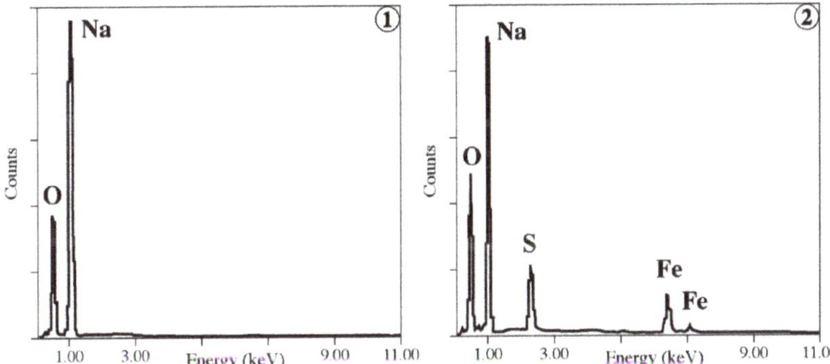

Figure 6. Scanning electron microscope coupled with energy dispersive spectrometry (SEM-EDS) examination results of a NaOH pellet reacted with $FeSO_4 \cdot H_2O$: (1) core of the NaOH pellet; (2) outer part of the NaOH pellet.

The examination of these results suggests the following deductions:

- The external part of the NaOH pellets (spectrum 2) is essentially composed of Na, O, S, and Fe, indicating the "reaction" of NaOH with iron sulfate,
- The NaOH pellets become porous (photo of Figure 6) facilitating the diffusion of reactive gases (chlorine) during the Na-ferrate synthesis,
- The core of the pellets is composed of Na and O showing the presence of unreacted NaOH (note that the amount of NaOH was 4 times more than the stoichiometry of the reaction: $2NaOH + FeSO_4 \rightarrow Fe(OH)_2 + Na_2SO_4$).

XRD analysis showed the presence of NaOH and $NaOH \cdot H_2O$ as predominant crystallized phases in the mixture. The iron-bearing phase was not revealed by XRD. However, the chemical analysis of the obtained mixture indicated that this mixture contained about 93 g Fe/kg and that 54% of iron was in a three-valence state. The presence of Fe(III) was also confirmed by Mössbauer spectroscopy measurements.

These pre-mixing materials were subjected to the Na-ferrate synthesis in a fluidized bed using diluted chlorine as an oxidation reagent.

3.4. Preliminary Fluidization Tests of the Premixing Materials in the Fluidized Bed

As it is well known, the fluidization of a solid by gas depends on the physical characteristics of the solid such as the density, particle size, shape, and those of the gas (gas viscosity and density).

The Reynolds (Re_{MF}) and Archimedes (Ar) numbers as well as the minimum velocity of fluidization (U_{MF}) are calculated by using the relationships available in the literature. The correlation of Wen and Yu [31] is used in this work, which is considered as the most known relationship related to a narrow particle size distribution. Equations (6)–(8) describe the formulas for calculating Re_{MF}, Ar, and U_{MF}.

Reynolds number at minimum fluidization (Re_{MF}):

$$Re_{MF} = \left[33.7^2 + 0.0408 \frac{d_p^3 \rho_g (\rho_s - \rho_g) g}{\mu_g^2} \right]^{1/2} - 33.7 \tag{6}$$

where μ_g is gas viscosity (Pa·s(kg·s^{-1}·m^{-1})); ρ_g is gas density (kg·m^{-3}); ρ_s is particle density (kg·m^{-3}); d_p is particle size (m); and g is gravitational constant (m·s^{-2}).

Archimedes number (Ar):

$$Ar = \frac{d_p^3 \rho_g (\rho_s - \rho_g) g}{\mu_g^2}. \tag{7}$$

Minimum velocity of fluidization (U_{MF}):

$$U_{MF} = Re_{MF} \frac{\mu_g}{d_p \rho_g} \tag{8}$$

The numeric substitution of the (μ_g, ρ_g, ρ_s, d_p, g) values showed that the mean minimum velocity of the fluidization of the premixing materials is about 1 m·s^{-1}. However, these values are approximate ones. For example, the solid density is considered to be 2130 kg·m^{-3} (density of NaOH), but it will be less if we consider that the NaOH particles become porous during the reaction with iron sulfate (see photo of Figure 6).

Based on these calculations, an FB of an internal diameter of about 1.75 cm was designed and constructed for the synthesis of Na-ferrate by using the NaOH conditioned as 2 mm pellets.

For the Na-ferrate synthesis, about 10 g of the premixed solids were loaded in the FB, creating a column of 9 to 10 cm. A total gas flow rate of 1100 L/h was necessary to ensure the fluidization of solids. The chlorine content of the used air + Cl_2 and N_2 + Cl_2 gas mixtures was kept at 5.5%, whilst the synthesis time was fixed at 15 min. The temperature of the thermostated water varied from 25 to 55 °C.

Visual observations indicated a good fluidization of the solids in FB without dust formation. This reinforces the idea that iron sulfate was well cemented during premixing (NaOH + FeSO$_4$·H$_2$O) in the rotary reactor. The solids after treatment in FB were examined by VM. The images of VM are compared with those of NaOH and premixing solids as shown in Figure 7. The results confirmed that the solids before reaction with chlorine (Figure 7b) and after (Figure 7c) had similar obvious shapes. The purple color of the solid surfaces suggested the formation of Na-ferrate. Furthermore, the dissolution in water of the solids coming from FB gave evidence of iron presence in its hexavalent state. The experimental conditions, as well as the chemical analysis of products issued from the synthesis process, are summarized in Table 1.

The synthesis products contained about 10 to 21 g/kg of iron in a hexavalent state. The best results were obtained when nitrogen was used as diluting gas. This is probably due to the presence of moisture in the air, which can decompose the synthesized ferrate. A higher Fe(VI) yield was achieved when the water bath was regulated at 55 °C. These preliminary tests showed the possibility of the Na-ferrate synthesis in a fluidized bed using the mixture (NaOH + FeSO$_4$·H$_2$O) as raw materials.

However, as the NaOH-pellets of 2 mm are no longer found in the market, it is suggested to test the synthesis of Na-ferrate using NaOH-pearls of about 1 mm (1000 µm).

Figure 7. Visible microscopy images of NaOH-pellets at different steps of the process (the small graduation of the paper in the background is 1 mm large). (a) Initial state of NaOH; (b) NaOH + FeSO$_4$· H$_2$O (premixing); (c) NaOH + FeSO$_4$·H$_2$O + Cl$_2$ (fluidized bed).

Table 1. Experimental conditions and Fe(VI) efficiency of synthesis using NaOH pellets of 2 mm.

Gas Mixture, Flow Rate		Cl$_2$ (%)	T (°C)	t (min)	Fe(VI) (g/kg)	Fe(VI) (%)
Air + Cl$_2$, L/h	1100	≈5.5	25	15	14.8	20.1
	1100	≈5.5	35	15	15.8	21.2
N$_2$ + Cl$_2$, L/h	1100	≈5.5	25	15	9.8	12.7
	1100	≈5.5	35	15	14.4	17.6
	1100	≈5.5	45	15	18.4	22.8
	1100	≈5.5	55	15	21.1	28.8

3.5. Premixing of NaOH Pearls (1000 µm) with Iron Sulfate

The tests carried out with pearls of sodium hydroxide and monohydrated ferrous sulfate in a reactor of 2.5 L (see Figure 2) showed that these reagents did not react like the NaOH pellets of 2 mm. Therefore, it was suggested to add FeSO$_4$·7H$_2$O in order to initiate the reaction. One may note that the particle size of FeSO$_4$·7H$_2$O is about 0.7 mm.

Figure 8 is a typical example of measured temperatures versus time in case of using ferrous sulfate with 3.5 and 4 mol of water (mixture of FeSO$_4$·7H$_2$O and FeSO$_4$·H$_2$O).

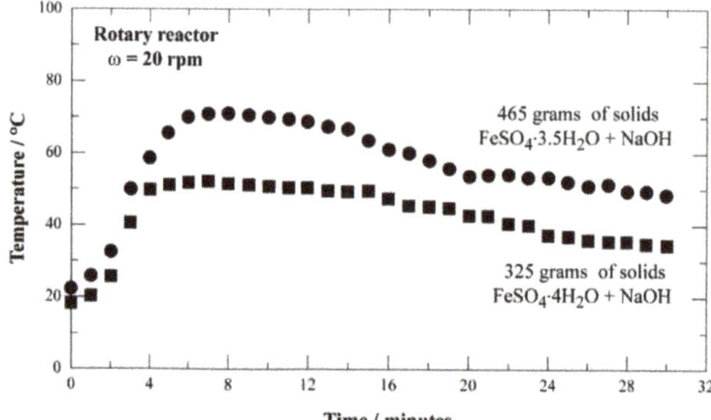

Figure 8. Evolution of the temperature as a function of time during the mixing of NaOH with iron sulfate (≈0.7 mm) containing 3.5 and 4 mol of water.

The molar ratio Na/Fe was fixed at 8 (to satisfy Equation (2) for the subsequent ferrate synthesis). This figure indicates that it was possible to mix about 325 and 465 g by using a hydration degree of 4 and 3.5 mol for the iron sulfate, respectively. The maximum temperature in the reactor did not exceed 80 °C. However, a close examination of the obtained mixture revealed that iron sulfate heptahydrate had not fully reacted with NaOH. This situation is presented in Figure 9 (visible microscope). Figure 9a showed the initial state of NaOH pearls, while Figure 9b represents the VM view of the mixture. As shown by Figure 9b, a large amount of $FeSO_4 \cdot 7H_2O$ is oxidized, agglomerated, and was difficult to separate by sieving.

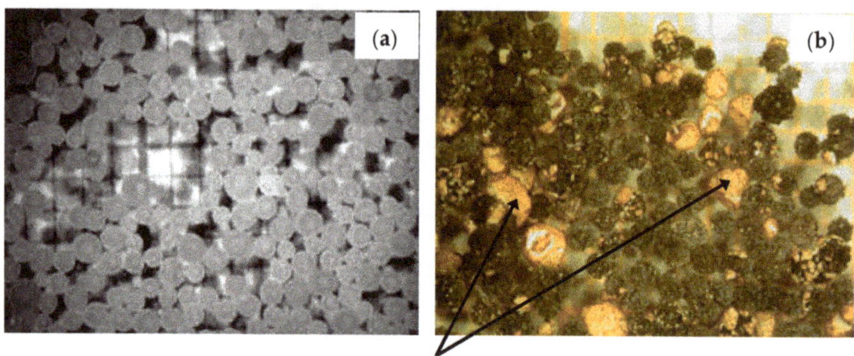

Iron sulfate oxidized and difficult to be separated

Figure 9. Visible microscopy images of sodium hydroxide (pearls-1 mm): (a) Initial state; (b) after reaction with $FeSO_4 \cdot 4H_2O$ (≈0.7 mm).

To overcome the phenomenon observed above, it was planned to use iron sulfate heptahydrate in powder form. The use of 0.12 mol $FeSO_4 \cdot 7H_2O$ in powder form, 0.88 mol $FeSO_4 \cdot H_2O$, and 8 mol NaOH provided the best results for obtaining the mixture (NaOH + $FeSO_4 \cdot 7H_2O$ + $FeSO_4 \cdot H_2O$). The results of several solid premixing tests are plotted in Figure 10, as the evolution of reactor temperature against reaction time in the case where about of 500 g of solids are used. As it could be expected, the reaction was exothermic, and the maximum temperature oscillated between 80 and 100 °C. The output mixture was sieved at +850 μm and was used for the synthesis of Na-ferrate in a fluidized bed.

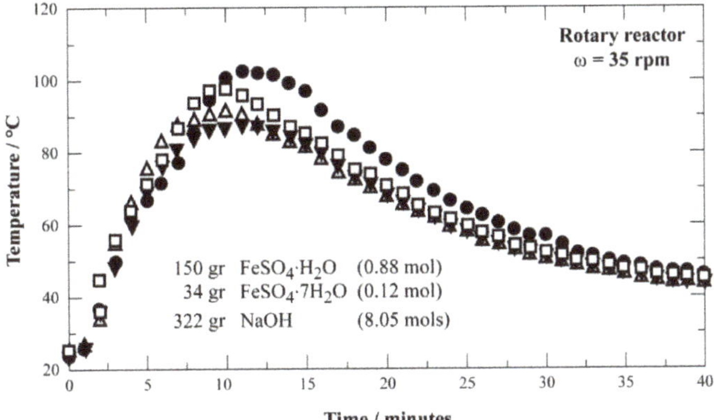

Figure 10. Evolution of the temperature as a function of time during mixing of NaOH with iron sulfate (powder) containing about 1.74 mol of water (four trials run in same conditions).

Figure 11a,b compares the VM images of the initial state of NaOH pearls and that of the obtained mixture according to trials mentioned above. The obtained mixture had a morphology similar to the initial NaOH particles.

(a)   (b)

Figure 11. VM views of sodium hydroxide at: (**a**) Initial state, ≤1 mm; (**b**) After reacting with powder iron sulfate.

3.6. Synthesis of Na-Ferrate in Various Fluidized Beds

A fluidized bed with a cross section of about 7.07 cm² ($\varnothing$ = 3 cm) was used for the synthesis of Na-ferrate based on the NaOH-pearls mixed previously with iron sulfate. It was interesting to follow the evolution of the temperature inside the fluidized bed as the reaction progressed; this was performed by placing a thermocouple in the fluidized bed and recording the temperature.

About 10 g of prepared mixture were loaded in the FB. Tests were performed at regulated water temperatures varying from 20 °C to 65 °C. The total flow rate of (air + Cl_2) and (N_2 + Cl_2) gaseous mixture was 1800 L/h corresponding to the operational fluidization velocity for the particles in the FB of $\varnothing$ 3 cm. The data obtained for a water temperature set at 35 °C are plotted in Figure 12a,b when air + Cl_2 and N_2 + Cl_2 are used, respectively. The chlorine contents of both gaseous mixtures were fixed at 0.6% and 2.2% Cl_2.

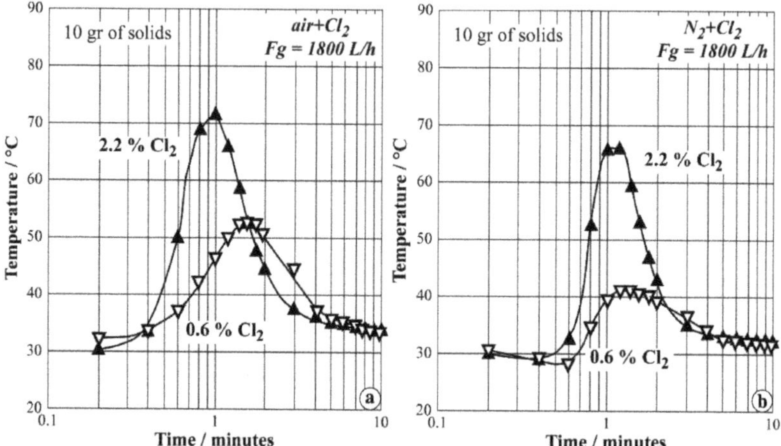

Figure 12. Plots of temperature evolution versus time during the synthesis of Na-ferrate in FB (Ø 3 cm) at 35 °C for: (**a**) air + Cl_2; (**b**) N_2 + Cl_2.

These figures clearly show that the temperature inside the FB increased sharply during the first minute of treatment. Use of 2.2% Cl_2 led to a higher maximum temperature level in both cases. The maximum temperatures observed were slightly lower when nitrogen was used instead of air. These observations are also valid over a water temperature range between 20 °C and 65 °C. The obtained products from these tests were examined by dissolving them in water. It was observed that most of the synthesized product did not show evidence of ferrate presence in the product. Data analyses suggested that the partial pressure of chlorine in the system was too low, making the oxidation of iron at high valence impossible. In order to have the flexibility of supplying high chlorine partial pressure, it was suggested to use a fluidized bed with a smaller diameter.

Tests in the FB with Ø 2 cm were carried out in a similar procedure as described for the previous case scenario. The total gas flow rate was 800 L/h corresponding to the operational fluidization velocity as in the case of FB with Ø 3 cm. The regulation temperature of water was varied from 18 °C to 55 °C, while the chlorine content was fixed at 2.5% and 5.0%. The evolution of the temperature as a function of the reaction time had a shape similar to that observed with FB Ø 3 cm.

The visual tests in water for the obtained product indicated that all the synthesis products at 2.5% chlorine were characterized by the absence of Na-ferrate. As a contrast, the synthesis products at 5.0% chlorine and the water regulated temperature higher than 18 °C showed some evidences of the presence of Na-ferrate. These results confirmed that the Na-ferrate synthesis is strongly affected by chlorine partial pressure.

Several synthesis products were subjected to a chemical analysis to determine the content in total iron and Fe(VI). A summary of the experimental conditions of the Na-ferrate synthesis in FB of Ø 2 cm and chemical analyses results is given in Table 2. These results indicated that the synthesis product (after 5 min of synthesis) contained between 72 and 74 g/kg of iron with about one half of iron as Fe(VI). Furthermore, the use of air instead of nitrogen decreased slightly the Fe(VI) efficiency.

Table 2. Experimental conditions and Fe(VI) efficiency of synthesis in FB of Ø = 2 cm.

Gas Mixture, Flow Rate	Cl_2 (%)	T (°C)	t (min)	Fe(VI) (g/kg)	Fe(VI) (%)
Air + Cl_2, L/h 800	≈5.0	40	5	36.2	49.3
800	≈5.0	50	5	36.8	50.3
N_2 + Cl_2, L/h 800	≈5.0	40	5	39.7	55.5
800	≈5.0	50	5	37.5	51.0

Although the obtained results for the Na-ferrate synthesis efficiency were improved in the smallest fluidized bed ($\varnothing$ = 2 cm), it was observed that the fluidization regime was better in the FB of $\varnothing$ = 3 cm. For this reason, the fluidized bed of intermediate diameter ($\varnothing$ = 2.5 cm) was also checked. The experimental procedures of the corresponding tests are similar to those developed for the FB of 3 and 2 cm.

Several experimental results are presented in Figure 13a–d plotted as the evolution of temperature inside the fluidized bed versus the reaction time in a half-logarithm scale. As in the previous case, the synthesis process seems to be completed in 5 min.

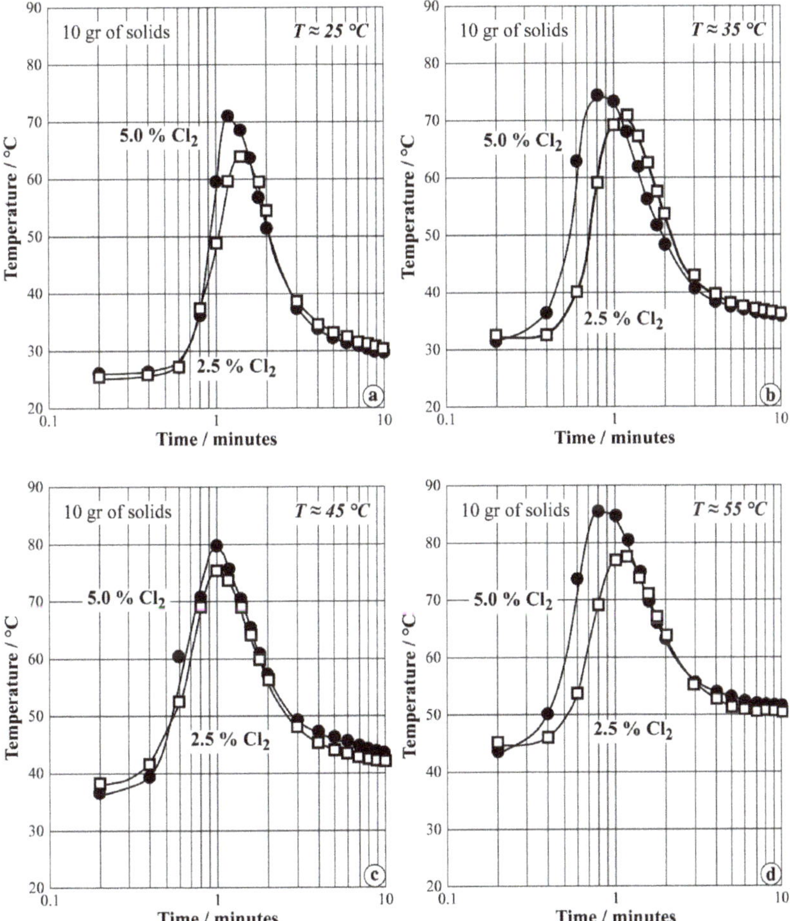

Figure 13. Plots of temperature evolution versus time during synthesis of Na-ferrate in FB (2.5 cm) using air + Cl_2 for regulated water temperature at: (**a**) 25 °C; (**b**) 35 °C; (**c**) 45 °C; (**d**) 55 °C.

Table 3 groups the experimental conditions and chemical analysis for three chosen tests of the whole series of experiments performed with FB of $\varnothing$ = 2.5 cm. The iron content of the synthesis process is about 75 g/kg, while the Fe(VI) yield varied between 43% and 54%. It seems that the Fe(VI) synthesis efficiency decreased when the regulated temperature exceeded 40 °C. The fluidization regime in FB of $\varnothing$ = 2.5 cm is better (more homogenous) than in the case of $\varnothing$ = 2 cm.

Table 3. Experimental conditions and Fe(VI) efficiency of synthesis in FB of $\varnothing = 2.5$ cm.

$N_2 + Cl_2$ (L/h)	Cl_2 (%)	T (°C)	t (min)	Fe_{Total} (g/kg)	Fe(VI) (g/kg)	Fe(VI) (%)
1200	≈5.0	30	5	74.6	40.1	53.8
1200	≈5.0	40	5	74.7	39.6	53.0
1200	≈5.0	50	5	74.4	31.9	42.9

3.7. Recycling of Non-Reacted Chlorine

In optimum conditions of the fluidized bed tests, the content of chlorine in the gaseous mixtures (air + Cl_2 and/or $N_2 + Cl_2$) was kept at 5%. As could be expected, only a certain fraction of chlorine is used for the ferrate synthesis. A part of unreacted chlorine was released through off-gases. Attempts were made to recycle this non-reacted chlorine. This was realized by connecting two fluidized beds in series. The off-gases of the first fluidized bed are used to supply the second FB.

About 10 g of the premixing material (single solid) were loaded in each FB. Gas mixture of $N_2 + Cl_2$ (5% Cl_2), which was introduced in the first FB, also passed through the second FB. A good fluidization of the first FB was observed, while the fluidization of second FB was more difficult, and the material showed a tendency to agglomerate. This is due to the water released during the ferrate synthesis in the first FB and causing a material agglomeration in the second FB. Therefore, it was suggested that the non-reacted chlorine can be recycled after a drying process to eliminate any trace of humidity. This can be easily achieved on an industrial scale, guaranteeing that no gaseous effluent is generated during the ferrate synthesis through the proposed process.

3.8. Environmental Considerations of the Process

Although this research is devoted to the preparation of alkali metal ferrates(VI) at room temperature, only some selected literature reports [32–43] prove the attractive properties of Fe(VI) for different end-use applications. Adjectives such as "environmentally friendly oxidant", "green oxidant", "strong oxidant", "powerful oxidant", and "super-iron battery" are often reserved for this class of materials.

The synthesis of ferrates, as described by this process, should be considered as a "green chemistry" process. The following paragraphs list some supporting evidence of this by comparing the principles of green chemistry [44] with our suggested process.

- Atom economy—synthetic methods should be designed to maximize the incorporation of all materials used in the process into the final product (almost all materials used are found in the final synthesis product);
- Safer solvents and auxiliaries—the use of auxiliary substances should be made unnecessary wherever possible and innocuous when used (no solvent or auxiliary substances—separation agents—are used for the ferrate synthesis by the proposed process);
- Design for energy efficiency—if possible, synthetic methods should be conducted at ambient temperature and pressure (it is exactly the case of the ferrate synthesis by the proposed process);
- Use of renewable feedstocks—a raw material or feedstock should be renewable rather than depleting whenever technically and economically practicable (raw materials used for the ferrate synthesis, i.e., Cl_2 and NaOH supply, is considered as almost as non-depletable);
- Design for degradation—chemical products should be designed so that at the end of their function, they break down into innocuous degradation products and do not persist in the environment (ferrates are transformed during their usage into hydroxides and/or oxides of Fe(III), considered as environmentally friendly compounds).

4. Conclusions

In this research, the possibility of transforming an industrial waste into supper-oxidizing materials (alkali metal ferrates) containing iron in its hexavalent state is shown. The following conclusions may be drawn from this investigation:

The proposed process for the synthesis of alkali ferrates included two main steps: (i) Premixing of NaOH with iron sulfate (solid–solid reactions) leading to a single solid, and (ii) fluidization of the obtained mixture in diluted chlorine (gas–solid reactions).

The premixing of the solids was achieved in a rotary reactor and the overall reaction was exothermic. The number of solids, the NaOH particle size, the Na/Fe ratio, and the moisture of the input materials ($FeSO_4 \cdot H_2O$ and/or $FeSO_4 \cdot 7H_2O$) affected this step. Temperatures lower than 100 °C were required to obtain good results for the premixing.

The output materials of the premixing, a single solid, contains about 80 to 90 g of Fe/kg of iron depending on the experimental conditions. Both Fe(II) and Fe(III) are present in the obtained mixture.

Various fluidized beds (with different cross sections) were designed and tested for the fluidization of the already prepared mixture. The fluidization was realized by air + Cl_2 and/or N_2 + Cl_2 assuring operational fluidization velocities.

The synthesis tests are carried out by varying parameters such as temperature of the thermostated water, partial pressure of chlorine, type of NaOH (2-mm pellets and 1-mm pearls), reaction time, etc.

The results suggest that fluidization is easy to achieve and that almost no dust is generated during the Na-ferrate synthesis in the fluidized bed, indicating that the iron sulfate is well embedded in the NaOH grains during the premixing step. The synthesis process is exothermic, and it is completed within a few minutes. Heat and water are rapidly evacuated from the reaction zone, leading to a dried Na-ferrate. This is a substantial advantage of performing the ferrate synthesis in a fluidized bed.

It seems that temperatures regulated close to 30 °C and temperatures in the fluidized bed lower than or equal to 70 °C (due to the exothermic reactions) provide the best results for the Na-ferrate synthesis. The Fe(VI) synthesis efficiency varied between 30% and 55% depending on other experimental parameters.

Ferrates obtained by this process could be used directly for different applications without any additional preliminary treatment (such as crushing), keeping in mind that the mean particle size of the ferrate produced by this invented process is close to 1 mm.

The proposed synthesis process meets numerous green chemistry and sustainable development principles. The synthesis product, alkali metal ferrates(VI), belongs to an advanced materials category with multipurpose functions for water and wastewater treatment as well as for cleaning various industrial effluents.

5. Patents

Kanari, N. Method of producing ferrates(VI). French patent, publication date: 14 March 2008, no. 2 905 609. Extension at international level: 13 March 2008, no. WO 2008/029046.

Author Contributions: Conceptualization, N.K.; formal analysis, C.D., I.F., S.S. and F.D.; investigation, N.K., E.O., C.D. and F.D.; visualization, E.O., I.F., S.S. and F.D.; resources, N.K., C.D. and J.Y.; writing—original draft, N.K., E.O., E.A. and F.P.; writing—review and editing, N.K., E.A., F.P. and J.Y.

Funding: A significant amount of this work is performed in the frame of contract no. G5RD-CT-2001-03011 (FP5-GROWTH) of the European Union. Further developments of the work have been supported by ANR program "Investissements d'avenir"—ANR-10-LABX-21-01/LABEX RESSOURCES21.

Conflicts of Interest: The authors declare no conflict of interest.

References

1. Kanari, N.; Menad, N.-E.; Ostrosi, E.; Shallari, S.; Diot, F.; Allain, E.; Yvon, J. Thermal behavior of hydrated iron sulfate in various atmospheres. *Metals* **2018**, *8*, 1084. [CrossRef]
2. Kanari, N.; Filippova, I.; Diot, F.; Mochón, J.; Ruiz-Bustinza, I.; Allain, E.; Yvon, J. Utilization of a waste from titanium oxide industry for the synthesis of sodium ferrate by gas-solid reactions. *Thermochim. Acta* **2014**, *575*, 219–225. [CrossRef]
3. Council Directive 92/112/EEC of 15 December 1992 on Procedures for Harmonizing the Programmes for the Reduction and Eventual Elimination of Pollution Caused by Waste from the Titanium Dioxide Industry.

Available online: https://eur-lex.europa.eu/legal-content/EN/TXT/?uri=CELEX%3A31992L0112 (accessed on 18 May 2019).

4. Gázquez, M.J.; Bolívar, J.P.; García-Tenorio, R.; Vaca, F. Physicochemical characterization of raw materials and co-products from the titanium dioxide industry. *J. Hazard. Mater.* **2009**, *166*, 1429–1440. [CrossRef] [PubMed]

5. Li, X.; Lei, Z.; Qu, J.; Li, Z.; Zhou, X.; Zhang, Q. Synthesizing slow-release fertilizers via mechanochemical processing for potentially recycling the waste ferrous sulfate from titanium dioxide production. *J. Environ. Manag.* **2017**, *186*, 120–126. [CrossRef] [PubMed]

6. Wu, L.; Wang, Z.; Li, X.; Guo, H.; Li, L.; Wang, X.; Zheng, J. Cation-substituted LiFePO$_4$ prepared from the FeSO$_4$·7H$_2$O waste slag as a potential Li battery cathode material. *J. Alloys Compd.* **2010**, *497*, 278–284. [CrossRef]

7. Huang, P.; Deng, S.; Zhang, Z.; Wang, X.; Chen, X.; Yang, X.; Yang, L. A sustainable process to utilize ferrous sulfate waste from titanium oxide industry by reductive decomposition reaction with pyrite. *Thermochim. Acta* **2015**, *620*, 18–27. [CrossRef]

8. Kanari, N.; Evrard, O.; Neveux, N.; Ninane, L. Recycling ferrous sulfate via super-oxidant synthesis. *JOM* **2001**, *53*, 32–33. [CrossRef]

9. Kanari, N.; Ostrosi, E.; Ninane, L.; Neveux, N.; Evrard, O. Synthesizing alkali ferrates using a waste as a raw material. *JOM* **2005**, *57*, 39–42. [CrossRef]

10. Kanari, N.; Filippov, L.; Diot, F.; Mochón, J.; Ruiz-Bustinza, I.; Allain, E.; Yvon, J. Synthesis of potassium ferrate using residual ferrous sulfate as iron bearing material. *J. Phys. Conf. Ser.* **2013**, *416*, 012013. [CrossRef]

11. Fremy, E. Recherches sur l'action des peroxides alcalins sur les oxides métalliques: Lettre de M. ED. Fremy à M. Pelouze. *Comptes Rendus de l'Académie des Sci. Paris* **1841**, *12*, 23–24.

12. Fremy, E. Recherches sur les acides métalliques. *Comptes Rendus de l'Académie des Sci Paris* **1842**, *14*, 442–446.

13. Fremy, E. Recherches sur les acides métalliques. *Annales de Chimie et de Phys.* **1844**, 361–382.

14. Whal, K.; Klemm, W.; Wehrmeyer, G. Über einige oxokomplexe von übergangselementen. *Z. Anorg. Allg. Chem.* **1956**, *285*, 322–336. [CrossRef]

15. Kiselev, N.S.; Kopelev, N.S.; Zavyalova, N.A.; Perfiliev, Y.D. The preparation of alkali metal ferrates(VI). *Russ. J. Inorg. Chem.* **1989**, *34*, 1250–1253.

16. Cici, M.; Cuci, Y. Production of some coagulant materials from galvanizing workshop waste. *Waste Manag.* **1998**, *17*, 407–410. [CrossRef]

17. Ockerman, L.T.; Schreyer, J.M. Preparation of sodium ferrate (VI). *J. Am. Chem. Soc.* **1951**, *73*, 5479. [CrossRef]

18. Schreyer, J.M.; Thompson, G.W.; Ockerman, L.T. Potassium ferrate (VI). *Inorg. Synth.* **1953**, *4*, 164–167.

19. Williams, D.H.; Riley, J.T. Preparation and alcohol oxidation studies of the ferrate (VI) ion. *Inorg. Chim. Acta* **1974**, *8*, 177–183. [CrossRef]

20. Lee, Y.; Cho, M.; Kim, J.Y.; Yoon, J. Chemistry of ferrrate (Fe(VI)) in aqueous solution and its applications as a green chemical. *J. Ind. Eng. Chem.* **2004**, *10*, 161–171.

21. Licht, S.; Tel-Vered, R.; Halperin, L. Direct electrochemical preparation of solid Fe(VI) ferrate, and super-iron battery compounds. *Electrochem. Commun.* **2002**, *4*, 933–937. [CrossRef]

22. Mácová, Z.; Bouzek, K.; Híveš, J.; Sharma, V.K.; Terryn, R.J.; Baum, J.C. Research progress in the electrochemical synthesis of ferrate(VI). *Electrochim. Acta* **2009**, *54*, 2673–2683. [CrossRef]

23. Sun, X.; Zhang, Q.; Liang, H.; Ying, L.; Xiangxu, M.; Sharma, V.K. Ferrate(VI) as a greener oxidant: Electrochemical generation and treatment of phenol. *J. Hazard. Mater.* **2016**, *319*, 130–136. [CrossRef] [PubMed]

24. Kanari, N.; Allain, E.; Shallari, S.; Diot, F.; Diliberto, S.; Patisson, F.; Yvon, J. Thermochemical route for extraction and recycling of critical, strategic and high value elements from by-products and end-of-life materials, Part I: Treatment of a copper by-product in air atmosphere. *Materials* **2019**, *12*, 1625. [CrossRef] [PubMed]

25. Jeannot, C.; Malaman, B.; Gérardin, R.; Oulladiaf, B. Synthesis, crystal and magnetic structures of the sodium ferrate (IV) Na$_4$FeO$_4$ studied by neutron diffraction and Mössbauer techniques. *J. Solid State Chem.* **2002**, *165*, 266–277. [CrossRef]

26. Kanari, N.; Allain, E.; Gaballah, I. Reactions of wüstite and hematite with different chlorinating agents. *Thermochim. Acta* **1999**, *335*, 79–86. [CrossRef]

27. Gaballah, I.; Ivanaj, S.; Kanari, N. Kinetics of chlorination and oxychlorination of chromium (III) oxide. *Metall. Mat. Trans. B* **1998**, *29A*, 1299–1308. [CrossRef]

28. Kanari, N.; Gaballah, I.; Allain, E. Kinetics of oxychlorination of chromite Part I. Effect of temperature. *Thermochim. Acta* **2001**, *371*, 143–154. [CrossRef]

29. Kanari, N.; Gaballah, I.; Allain, E. Kinetics of oxychlorination of chromite Part II. Effect of reactive gases. *Thermochim. Acta* **2001**, *371*, 75–86. [CrossRef]

30. Roine, A. *Outokumpu HSC Chemistry for Windows Version 3.0*; Outokumpu Research: Pori, Finland, 1997.

31. Wen, C.Y.; Yu, Y.H. A generalized method for predicting the minimum fluidization velocity. *AIChE J.* **1966**, *12*, 610–612. [CrossRef]

32. Sharma, V.K. Potassium ferrate(VI): An environmentally friendly oxidant. *Adv. Environ. Res.* **2002**, *6*, 143–156. [CrossRef]

33. Sharma, V.K. Ferrate(V) oxidation of pollutants: A premix pulse radiolysis study. *Radiat. Phys. Chem.* **2002**, *65*, 349–355. [CrossRef]

34. Ninane, L.; Kanari, N.; Criado, C.; Jeannot, C.; Evrard, O.; Neveux, N. New processes for alkali ferrate synthesis. In *Ferrates: Synthesis, Properties and Applications in Water and Wastewater Treatment*; Sharma, V.K., Ed.; American Chemical Society: Washington, DC, USA, 2008; pp. 102–111.

35. Jiang, J.Q. Research progress in the use of ferrate(VI) for the environmental remediation. *J. Hazard. Mater.* **2007**, *146*, 617–623. [CrossRef] [PubMed]

36. Alsheyab, M.; Jiang, J.-Q.; Stanford, C. On-line production of ferrate with an electrochemical method and its potential application for wastewater treatment—A review. *J. Environ. Manag.* **2009**, *90*, 1350–1356. [CrossRef] [PubMed]

37. Wang, Y.L.; Ye, S.H.; Wang, Y.Y.; Cao, J.S.; Wu, F. Structural and electrochemical properties of a K_2FeO_4 cathode for rechargeable Li ion batteries. *Electrochim. Acta* **2009**, *54*, 4131–4135. [CrossRef]

38. Filip, J.; Yngard, R.A.; Siskova, K.; Marusak, Z.; Ettler, V.; Sajdl, P.; Sharma, V.K.; Zboril, R. Mechanisms and efficiency of the simultaneous removal of metals and cyanides by using ferrate(VI): Crucial roles of nanocrystalline iron (III) oxyhydroxides and metal carbonates. *Chem. Eur. J.* **2011**, *17*, 10097–10105. [CrossRef]

39. Yates, B.J.; Zboril, R.; Sharma, V.K. Engineering aspects of ferrate in water and wastewater treatment—A review. *J. Environ. Sci. Health Part A* **2014**, *49*, 1603–1614. [CrossRef]

40. Jiang, Y.; Goodwill, J.E.; Tobiason, J.E.; Reckhow, D.A. Effect of different solutes, natural organic matter, and particulate Fe(III) on ferrate(VI) decomposition in aqueous solutions. *Environ. Sci. Technol.* **2015**, *49*, 2841–2848. [CrossRef]

41. Wei, Y.-L.; Wang, Y.-S.; Liu, C.-H. Preparation of potassium ferrate from spent steel pickling liquid. *Metals* **2015**, *5*, 1770–1787. [CrossRef]

42. Goodwill, J.E.; Jiang, Y.; Reckhow, D.A.; Tobiason, J.E. Laboratory assessment of ferrate for drinking water treatment. *J. Am. Water Works Assoc.* **2016**, *108*, E164–E174. [CrossRef]

43. Han, H.; Li, J.; Ge, Q.; Wang, Y.; Chen, Y.; Wang, B. Green ferrate(VI) for multiple treatments of fracturing wastewater: Demulsification, visbreaking, and chemical oxygen demand removal. *Int. J. Mol. Sci.* **2019**, *20*, 1857. [CrossRef]

44. Anastas, P.T.; Warner, J.C. *Green Chemistry: Theory and Practice*; Oxford University Press: New York, NY, USA, 1998; p. 30.

 © 2019 by the authors. Licensee MDPI, Basel, Switzerland. This article is an open access article distributed under the terms and conditions of the Creative Commons Attribution (CC BY) license (http://creativecommons.org/licenses/by/4.0/).

Article

Reactivity of Low-Grade Chromite Concentrates towards Chlorinating Atmospheres

Ndue Kanari [1,*], Eric Allain [1], Lev Filippov [1], Seit Shallari [2], Frédéric Diot [1] and Fabrice Patisson [3]

[1] Université de Lorraine, CNRS, GeoRessources, F-54000 Nancy, France; ericgallain@gmail.com (E.A.); lev.filippov@univ-lorraine.fr (L.F.); frederic.diot@univ-lorraine.fr (F.D.)

[2] Faculty of Agriculture and Environment, Agricultural University of Tirana, 1029 Tirana, Albania; seitshallari@gmail.com

[3] Université de Lorraine, CNRS, Labex DAMAS, IJL, F-54000 Nancy, France; fabrice.patisson@univ-lorraine.fr

* Correspondence: ndue.kanari@univ-lorraine.fr; Tel.: +33-372-744-530

Received: 28 August 2020; Accepted: 7 October 2020; Published: 9 October 2020

Abstract: The most economically important iron-chromium bearing minerals is chromite. In natural deposits, iron(II) is frequently substituted by magnesium(II) while chromium(III) is replaced by aluminum(III) and/or iron(III) forming a complex chromium bearing material. The majority of mined chromite is intended for the production of ferrochrome which requires a chromite concentrate with high chromium-to-iron ratio. Found mostly in the spinel chromite structure, iron cannot be removed by physical mineral processing methods. In this frame, the present work deals with the reaction of chlorine and chlorine+oxygen with selected samples of chromite concentrates for assessing the reactivity of their components towards chlorinating atmosphere, allowing the preferential removal of iron, hence meeting the chromite metallurgical grade requirements. Isothermal thermogravimetric analysis was used as a reliable approach for the kinetic reactivity investigation. Results indicated a wide difference in the thermal behavior of chromite constituents in a chlorinating atmosphere when considering their respective values of apparent activation energy oscillating from about 60 to 300 kJ/mol as a function of the sample reacted fraction. During the chromite treatment by chlorine in presence of oxygen, chromium was recovered as liquid chromyl chloride by condensation of the reaction gas phase.

Keywords: chromite; chlorine; thermogravimetric analysis; isothermal treatment; apparent activation energy

---

1. Introduction

Chromium is part of the extended group of refractory metals offering beneficial properties for various end-uses and manufacturing applications. As displayed in Figure 1, part of these metals (Nb, W, V, Hf, Ta) belongs to the critical materials, according to the European Union criticality assessment [1]. Although chromium is in the cut-off level to be critical, it ranks third (after Mg and W) from the standpoint of economic importance.

Most refractory metals are found and extracted from their oxide bearing materials, likewise, the only economic source of chromium is chromite ($FeCr_2O_4$) ore. Nevertheless, in natural deposits, Mg(II) may substitute Fe(II), while Al(III) and Fe(III) often substitute Cr(III) resulting to a complex chromium bearing mineral with the fairly general formula $(Fe,Mg)(Cr,Al,Fe)_2O_4$, encompassing the main end-members such as $FeO·Fe_2O_3$ (magnetite), $FeO·Cr_2O_3$ (iron chromite), $MgO·Cr_2O_3$ (magnesiochromite), $MgO·Al_2O_3$ (true/regularly spinel). Being multiple and complete solid solution of the spinel group, the composition of chromite is no fixed, but varies largely and depends on the geographic and geochemical features of its deposits.

The major part of mined chromite goes to the ferrochrome (FeCr) manufacture [2–5] and in turn, the FeCr is intended to stainless steels and chromium bearing alloys production. It seems that chromium has no substitute for these industrial end-uses. According to available data [6], the average annual growth of the world stainless steels production is about 5.84% reaching 50.7 million metric tons in 2018. Selected reports [7–13] from numerous recent research works reported to the scientific journals *Materials* and *Metals* are focused on the chromium bearing steels and alloys showing the importance of these leading areas for chromium utilization driving the needs for the ferrochrome and chromite production. Chemical industry, foundry sands and refractory segments are other end-uses of chromite, but their weight relative to the total chromite demand is minor.

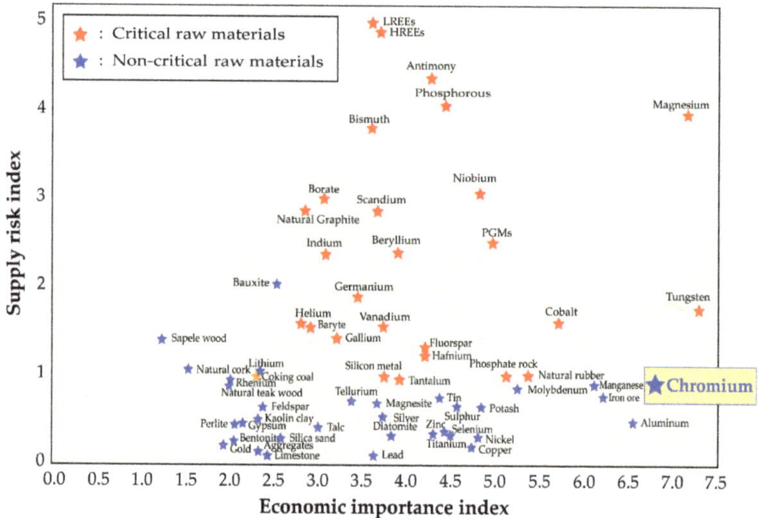

Figure 1. The 2017 criticality assessment of raw materials for the European Union.

It must be emphasized that the production of ferrochrome not only requires a chromite ore and/or concentrate with a high Cr_2O_3 content (46–48% Cr_2O_3), but also with a chromium-iron ratio above 2 (typically around 2.8). The physical processing of the mineral can be successful for the removal of the chromite ore gangue leading to a concentrate with a satisfactory Cr_2O_3 content meeting the metallurgical requirement. However, as the major part of iron is found in the lattice structure of chromite, only a chemical broken down of the chromite structure seems appropriate for the removal of iron from chromite. According to Nafziger [14], chemical techniques, such as hydrometallurgical methods, chlorination, roasting and leaching, as well as smelting, are required for increasing the chromium-to-iron ratio of lean chromite ores and concentrates. Recently, a carbo-thermic reduction followed by hydrochloric acid leaching was tried as an efficient method [15] for the extraction of iron from a poor chromium concentrate. Specific research works [16–21] regarding the use of various chlorination agents for chlorination of chromite ores and concentrate and their constituents were summarized earlier [2,3].

Our various research reports published previously [2,3,22–24] were focused on the carbochlorination, chlorination and oxychlorination of rich chromite concentrates having a chromium-to-iron ratio of around 3.2 suitable for ferrochrome manufacturing. However, no recent studies were disclosed in the literature regarding to the use of Cl_2+O_2 for processing lean chromite materials, i.e., having typically a chromium to iron ratio lesser than 2.8. In this regard, the present paper essentially describes the behavior of a poor chromite concentrate with a chromium-to-iron ratio near to 1.5, under Cl_2 and Cl_2+air atmospheres. The reactivity and behavior of chromite constituents were examined using thermogravimetric analysis (TGA) approach under isothermal conditions. For a better understanding of the processes, the experimental results are compared with those obtained for the treatment of a rich chromite concentrate (of metallurgical grade).

2. Materials and Methods

The first chromite concentrate sample (low grade) used for this investigation was provided by a European Union manufacturer. A second sample (high grade) provided by an Albanian chromite plant (Bulqiza, Albania) was also used, mostly for comparison purpose. The physico-chemical characterization of the chromite samples and of the reaction products was performed by diverse analytical methods such as chemical analysis (by inductively coupled plasma atomic emission spectroscopy "ICP-AES"), scanning electron microscopy-energy dispersive spectroscopy (SEM-EDS, HITACHI S-4800, Hitachi Ltd., Tokyo, Japan) and X-ray diffraction (XRD, Bruker D8 Advance device, Bruker, Karlsruhe, Germany). Their description was given in previous research works [2,25,26]; only the results will be reported hereafter.

Experimental tests of the reaction of chromite concentrates with chlorine were carried out in a vertical microbalance (model CAHN 1000, Cahn Co., Cerritos, CA, USA) operating at a sensitivity of 10 μg and designed to work under corrosive atmosphere. The equipment configuration with the accessory parts is shown schematically in Figure 2. Several milligrams (more often 40 to 50 mg) of sample were spread out in a silica crucible and the whole specimen was heated up to the desired temperature under nitrogen atmosphere. Subsequently, the nitrogen was replaced by Cl_2 and/or Cl_2+air (O_2) and the evolution of the mass loss over time was recorded. The accessory parts of the setup illustrated in Figure 2 are the units for measuring and purifying the inlet gases as well as those for neutralizing the outlet gases.

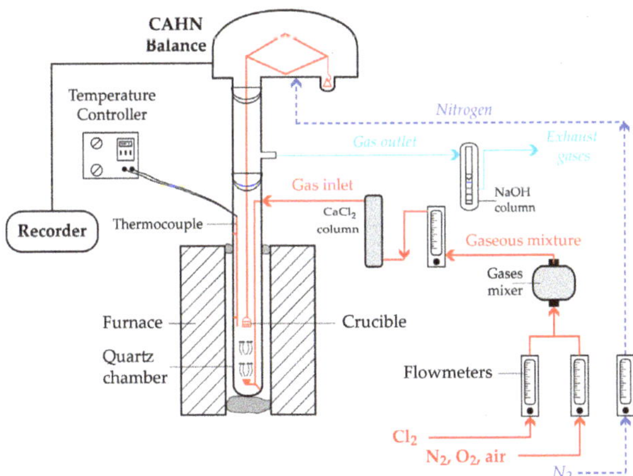

Figure 2. Setup of the TG analysis experiment.

3. Results

3.1. Physico-Chemical Characterization of Chromite Concentrate Samples

The chemical composition in the five main constituents of the first chromite sample is given in Figure 3. As shown, this concentrate is characterized by a high iron content (26.9% wt expressed as FeO) and its chromium-to-iron ratio of 1.48 makes it unsuitable for the FeCr manufacturing. This sample is denoted as low-grade chromite concentrate (LGChC). The XRD patterns of the sample matched well with the $(Fe,Mg)(Cr,Fe,Al)_2O_4$ phase. Note that the simple constituents of chromite

(Fe_3O_4, $FeCr_2O_4$, $MgCr_2O_4$ and $MgAl_2O_4$) are isomorphs having analogous XRD profile making their individual identification difficult. Based on the chemical analysis and supposing a perfect stoichiometric composition without cation/anion deficiency and/or defect, the general formula of the chromite body of LGChC can be approximately represented as $(Fe_{0.50},Mg_{0.50})(Cr_{1.20},Al_{0.60},Fe_{0.20})O_4$ and the chromite body can also be expressed as 15.1% Fe_3O_4, 37.6% $FeCr_2O_4$, 26.4% $MgCr_2O_4$ and 20.8% $MgAl_2O_4$ (Figure 4).

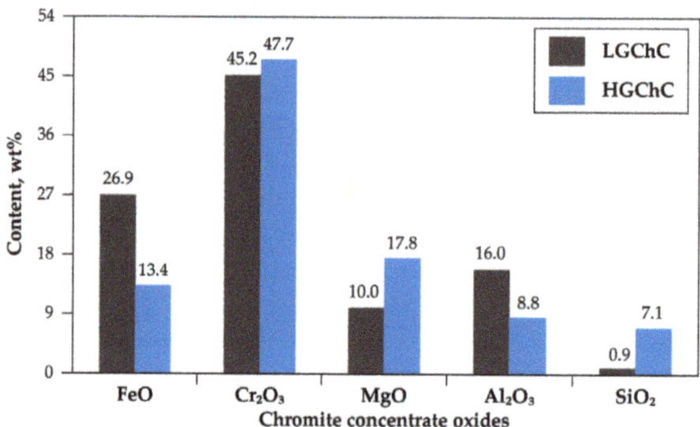

Figure 3. Chemical composition of the chromite concentrate samples.

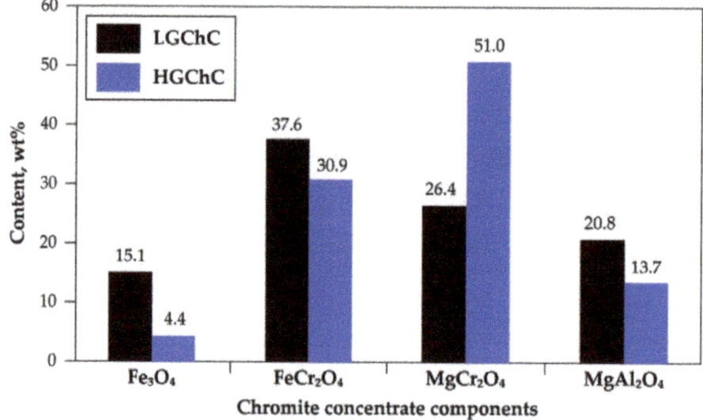

Figure 4. Mineralogical composition of the chromite concentrate samples.

The second sample had a chromium-to-iron ratio close to 3.2 and is labelled as high-grade chromite concentrate (HGChC). Although a rich concentrate with 47.7% Cr_2O_3 (Figure 3), it contained high amount of gangue (close to 7% SiO_2 belonging also to olivine and serpentine minerals). The chromite body was separated from the gangue by successive physical separations (using dense liquor) and was defined as $(Fe_{0.30}, Mg_{0.70})(Cr_{1.56},Al_{0.37},Fe_{0.07})O_4$ with the average composition of end-members 4.4% Fe_3O_4, 30.9% $FeCr_2O_4$, 51.0% $MgCr_2O_4$ and 13.7% $MgAl_2O_4$ (Figure 4). The mean particle size of both chromite samples used for this study is less than 100 μm. As the reactions of HGChC with Cl_2+CO, Cl_2 and Cl_2+O_2 gaseous mixtures were studied thoroughly in previous research work [2,3,22–24], its behaviour in the chlorinating atmosphere is used as a reference to explain the phenomena observed during the treatment of the low-grade chromite concentrate.

3.2. Behavior of Chromite under Chlorine Atmosphere

Envisaged reactions of the complex chromite constituents (Fe_3O_4, $FeCr_2O_4$, $MgCr_2O_4$ and $MgAl_2O_4$) and those of simple oxides (FeO, Fe_2O_3, Cr_2O_3, MgO and Al_2O_3) with chlorine may be represented by Equation (1) through (9). The value of standard free energy changes ($\Delta_rG°$) at 900 °C is computed from HSC thermochemical database [27] and is indicated beside each reaction. According to these values, the reactions of chromite constituents with $Cl_{2(g)}$ (Equations (1)–(4)) proceed with $\Delta_rG° > 0$ indicating a nonspontaneous process in the forward direction; these reactions will absorb energy from its surroundings in order to take place. Among the reactions of simple metals oxides of the chromite with chlorine involving their respective chlorides (Equations (5) to (9)), only the reaction of FeO seems to be a spontaneous reaction from a thermodynamic point of view; Cr_2O_3 and Al_2O_3 are the most stable oxides in chlorine atmosphere.

$$\Delta_rG° \ (900 \ °C), \ kJ/mol \ Cl_2$$

$$2/9 \ Fe_3O_{4(s)} + Cl_{2(g)} \rightarrow 2/3 \ FeCl_{3(g)} + 4/9 \ O_{2(g)} \qquad 11.54 \qquad (1)$$

$$2/9 \ FeCr_2O_{4(s)} + Cl_{2(g)} \rightarrow 2/9 \ FeCl_{3(g)} + 4/9 \ CrCl_{3(s)} + 4/9 \ O_{2(g)} \qquad 54.03 \qquad (2)$$

$$1/4 \ MgCr_2O_{4(s)} + Cl_{2(g)} \rightarrow 1/4 \ MgCl_{2(l)} + 1/2 \ CrCl_{3(s)} + 1/2 \ O_{2(g)} \qquad 71.09 \qquad (3)$$

$$1/4 \ MgAl_2O_{4(s)} + Cl_{2(g)} \rightarrow 1/4 \ MgCl_{2(l)} + 1/2 \ AlCl_{3(g)} + 1/2 \ O_{2(g)} \qquad 78.47 \qquad (4)$$

$$2/3 \ FeO_{(s)} + Cl_{2(g)} \rightarrow 2/3 \ FeCl_{3(g)} + 1/3 \ O_{2(g)} \qquad -27.93 \qquad (5)$$

$$1/3 \ Fe_2O_{3(s)} + Cl_{2(g)} \rightarrow 2/3 \ FeCl_{3(g)} + 1/2 \ O_{2(g)} \qquad 19.53 \qquad (6)$$

$$1/3 \ Cr_2O_{3(s)} + Cl_{2(g)} \rightarrow 2/3 \ CrCl_{3(s)} + 1/2 \ O_{2(g)} \qquad 80.33 \qquad (7)$$

$$MgO_{(s)} + Cl_{2(g)} \rightarrow MgCl_{2(l)} + 1/2 \ O_{2(g)} \qquad 9.57 \qquad (8)$$

$$1/3 \ Al_2O_{3(s)} + Cl_{2(g)} \rightarrow 2/3 \ AlCl_{3(g)} + 1/2 \ O_{2(g)} \qquad 85.93 \qquad (9)$$

An important point of the thermodynamic findings is that the thermodynamic reactivity of the complex constituents of the chromite in chlorine is decreasing according to the following sequence:

$$Fe_3O_4 > FeCr_2O_4 > MgCr_2O_4 > MgAl_2O_4 \qquad (10)$$

The evolution of the vapor pressure of chlorides of main chromite elements (Cr, Fe, Mg, Al, Si) is displayed in Figure 5 [28,29]. It indicates that in the case of the chlorination of chromite constituents, a selective separation of the obtained chlorides is feasible thanks to great differences in their vapor pressure in a selected temperature interval. A special case is the high volatility of chromyl chloride (CrO_2Cl_2), which will be discussed in Section 3.3.

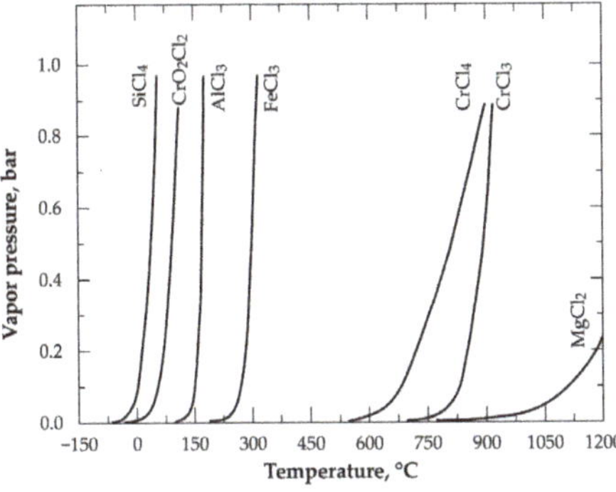

Figure 5. Vapor pressure versus temperature for several chlorides likely to be produced during chromite chlorination.

Based on these thermodynamic predictions and on the work previously performed [2], experimental tests of LGChC were conducted with chlorine alone with a flow rate of 60 L/h. The recorded data are depicted in Figure 6 as percent mass loss (% ML) of the sample over the reaction time. The somewhat atypical curve shape is a first indication of the complexity of the reactions of chromite with chlorine. As shown by Figure 6b, the first 50% of the sample was quickly chlorinated and volatilized, while the remaining sample appeared more refractory to chlorine. As an example, the isothermal data indicates that 32 min were sufficient to achieve 50% conversion at 950 °C while 75% conversion required 180 min at this temperature.

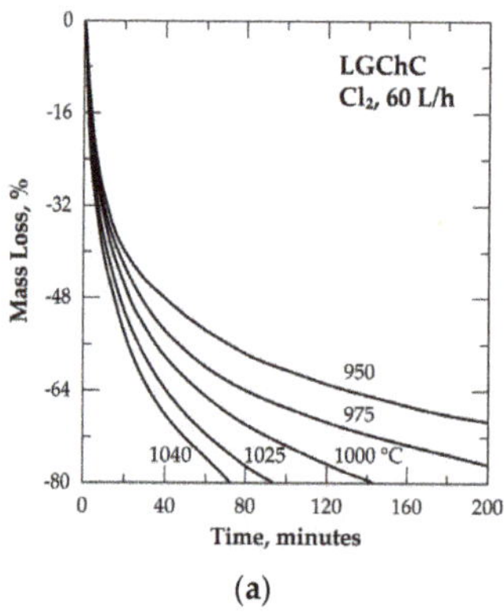
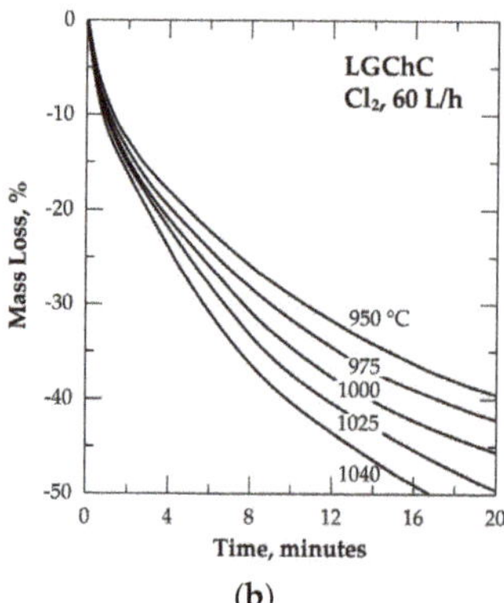

(a) **(b)**

Figure 6. Mass change of the sample versus time for the treatment of LGChC in chlorine from 950 to 1040 °C: (**a**) General view of the obtained isotherms; (**b**) Zoom in on the graph up to 50% ML.

To get an insight about this change in the curve shape, it is helpful to compare several isothermal TGA plots for both samples, i.e., LGChC and HGChC, as illustrated in Figure 7. The chlorination of chromite concentrates at 800 °C (Figure 7a) tends to an asymptote of the %ML beyond 2 h of treatment corresponding to about 50% and 35% ML for LGChC and HGChC, respectively.

Gathering this data with the mineralogical composition of chromite concentrates (Figure 4) and the thermodynamic predictions allows us to hypothesize that only magnetite and iron chromite are chlorinated at 800 °C. It was evaluated that these compounds ($Fe_3O_4+FeCr_2O_4$) represent close to 52.7% and 35.3% of the LGChC and HGChC, respectively. With this evidence, one may also conclude that the TG analysis of chromite reactions with chlorine at low temperature can be an effective method for the fair determination of the amount of ($Fe_3O_4+FeCr_2O_4$) contained in the chromite ores and/or concentrates. Although less wide, the difference between the %ML obtained for the LGChC and HGChC is still evident at 950 °C (Figure 7b) and 1040 °C (Figure 7c).

The direct application of well-known kinetics models [30] for describing the reaction progress and rate expression faces certain difficulties related to successive reactions, altering of the physical and chemical reactivity of the remaining sample, inter-reaction between gaseous reaction products and the working sample, etc. Hence, it was suggested that the best way to evaluate the temperature impact on the chromite reactions with Cl_2 was to calculate the reaction rate in increments of 5% mass losses in the interval ranging from 5.0% to 85.0% ML. This is performed for all isothermal data from 950 to 1040 °C. Shown in Figure 8 is the graphical representation of the processed data at 975 °C. Besides data linearization expressing the reaction rate, the correlation coefficient (R^2) of data fitting, for each segment of 5% ML, is also shown.

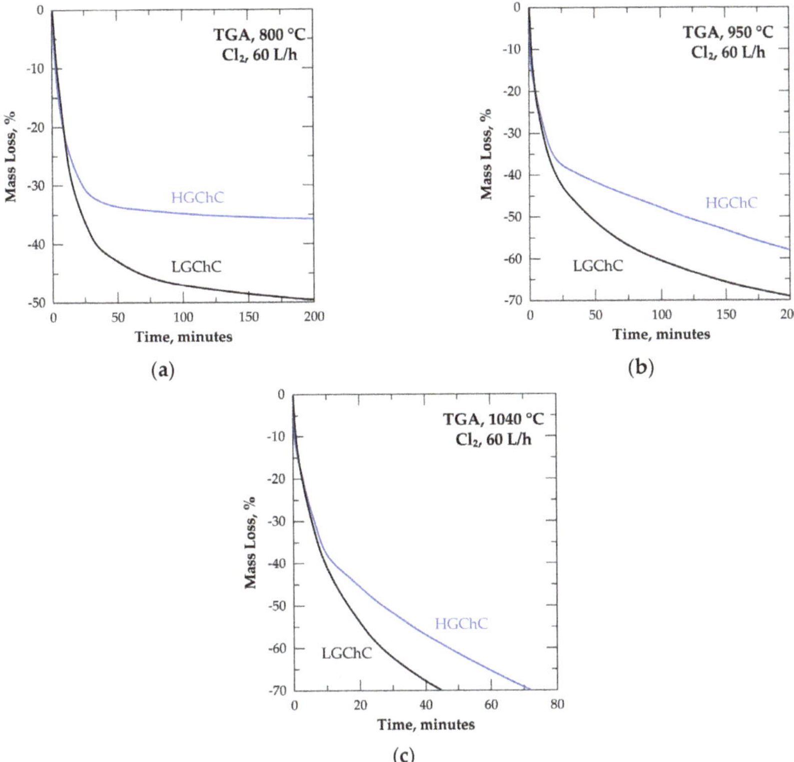

Figure 7. Comparison of the thermal behavior of LGChC and HGChC in Cl_2 atmosphere: (**a**) 800 °C; (**b**) 950 °C; (**c**) 1040 °C.

It was stated [2] that the reaction of chlorine with chromite concentrates generated metal chlorides having volatilization rate higher than their formation rate, indicating the %ML of the sample expresses directly the fraction of the sample (α) reacted. Therefore, the Arrhenius diagrams displaying the logarithm of the reaction rate plotted versus the inverse of the temperature for each 5% ML segment were drawn and values of the apparent activation energy (E_a) with standard error were computed.

Figure 9 gives Arrhenius' plots for the four chosen reacted fractions. Good fitting of the traced data is obtained with the value of E_a increasing from 58±5 kJ/mol at ($0.10 \leq \alpha \leq 0.15$) to 285 ± 8 kJ/mol at ($0.75 \leq \alpha \leq 0.80$) with a good confidence level. Furthermore, the reaction rate at 1000 °C for ($0.75 \leq \alpha \leq 0.80$) is decreased by around 27 times with respect to the reaction rate for ($0.10 \leq \alpha \leq 0.15$) at the same temperature, which seems unusual for the gas-solid reactions with gasification of reaction products and without formation of new solid products.

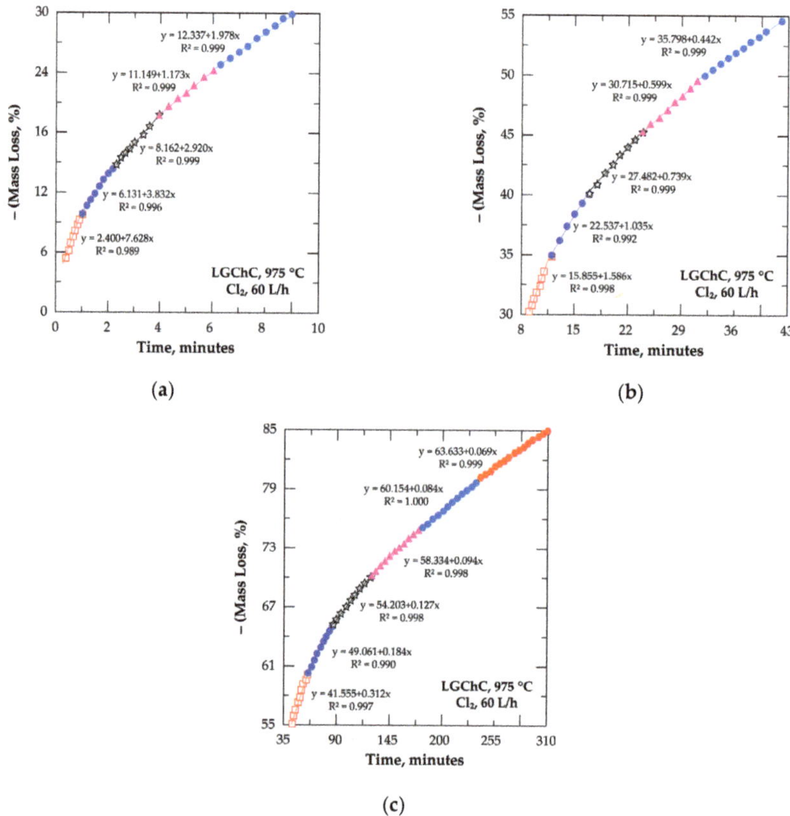

Figure 8. Plot of %ML versus time gathered with mean reaction rates calculated in increments of 5% ML during treatment of LGChC in chlorine at 975 °C: (**a**) 5–30% ML; (**b**) 30–55% ML; (**c**) 55–85% ML. The color markers are used to distinguish the segments of the %ML curves for which the linearization has been made for the calculation of the reaction rate.

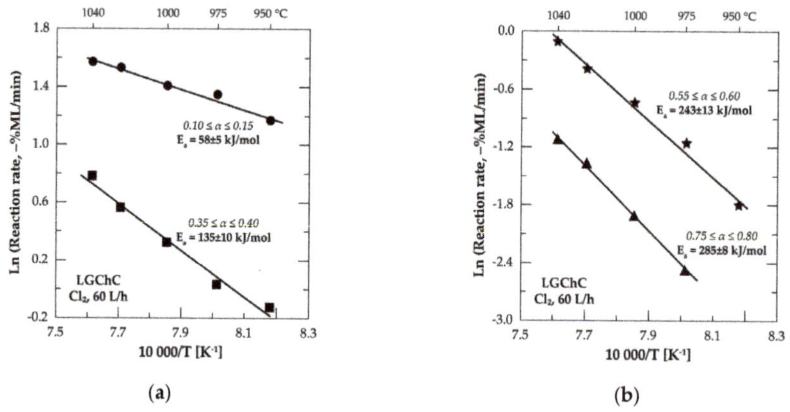

Figure 9. Examples of the Arrhenius diagrams for the reaction of LGChC with Cl_2: (**a**) $0.10 \leq \alpha \leq 0.15$ and $0.35 \leq \alpha \leq 0.40$; (**b**) $0.55 \leq \alpha \leq 0.60$ and $0.75 \leq \alpha \leq 0.80$.

An overall profile of the evolution of E_a as a function of the chromite conversion fraction is displayed in Figure 10. The beginning of the reaction proceeded with an E_a near to 78 kJ/mol and decreased to about 60 kJ/mol at $0.10 \leq \alpha \leq 0.20$. Such a trend suggests the formation of an intermediate species, unfortunately unknown, decreasing the potential barrier of the reaction. Based on the chemical and mineralogical composition of the LGChC the fraction converted at the beginning of reaction may be attributed to the reaction of Fe_3O_4 with Cl_2 involving ferric chloride as final product as it is highly volatile at this temperature range (Figure 5). Thereafter, an increase of the E_a is observed reaching about 175 kJ/mol at $(0.45 \leq \alpha \leq 0.55)$. One may attribute globally this value of E_a to the reaction of iron chromite ($FeCr_2O_4$) with chlorine. Next, at $\alpha > 0.60$, the apparent activation energy increased again reaching values as high as 304 kJ/mol for α ranging from 0.70 to 0.75. This conversion fraction corresponds most probably to the removal of $MgCr_2O_4$ from the chromite body. According to obtained apparent activation energy and reaction rate trends, the decreasing reaction rank of chromite constituents with chlorine follows the sequence:

$$Fe_3O_4 > FeCr_2O_4 > MgCr_2O_4 > MgAl_2O_4 \tag{11}$$

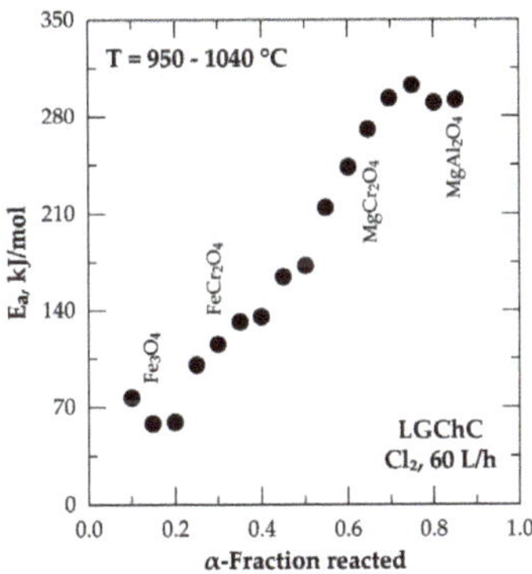

Figure 10. Plot of the apparent activation energy (E_a) versus fraction reacted for the treatment of LGChC in chlorine atmosphere between 950–1040 °C.

In other words, a higher reaction rate was associated with a lower value of E_a and vice versa. It is hence concluded that the reactivity of chromite constituents towards chlorine is in good agreement with the apparent activation energy and the reaction rates are sufficiently different to achieve a selective elimination of one constituent without affecting the other constituents. Another point is also to be mentioned that this reaction sequence (Equation (11)) seems to match well with the sequence based on thermodynamic predictions and shown in (Equation (10)).

3.3. Reactions of Chromite with Chlorine in Presence of Oxygen

Having obtained information on the reaction of chromite with chlorine, it was useful to investigate the impact of oxygen on the reaction kinetic and involved products. The chemical reactions of the two main chromite constituents, $FeCr_2O_4$ and $MgCr_2O_4$, with chlorine +oxygen can be described by Equations (12) and (13), respectively. The values of $\Delta_r G°$ are still positive, but they are much lower than those obtained for the chlorination with chlorine alone (Equations (2) and (3), respectively).

$$\Delta_r G° \ (900 \ °C), \ kJ/mol \ Cl_2$$

$$1/2 \ FeCr_2O_{4(s)} + Cl_{2(g)} + 3/8 \ O_{2(g)} \rightarrow 1/4 \ Fe_2O_{3(s)} + CrO_2Cl_{2(g)} \qquad 5.89 \qquad (12)$$

$$1/2 \ MgCr_2O_{4(s)} + Cl_{2(g)} + 1/4 \ O_{2(g)} \rightarrow 1/2 \ MgO_{(s)} + CrO_2Cl_{2(g)} \qquad 36.37 \qquad (13)$$

$$1/2 \ Cr_2O_{3(s)} + Cl_{2(g)} + 1/4 \ O_{2(g)} \rightarrow CrO_2Cl_{2(g)} \qquad 19.47 \qquad (14)$$

One reaction of particular interest in the system Cr-O-Cl is that of the chromium trioxide (Cr_2O_3) with chlorine in presence of oxygen with overall reaction described by Equation (14). As shown, the reaction consumes oxygen leading to the formation of CrO_2Cl_2 as final reaction product. The computed value of $\Delta_r G°$ (900 °C) is 19.47 kJ/mol instead of 80.33 kJ/mol for the chemical reaction of Cr_2O_3 with chlorine solely generating $CrCl_3$ at the same temperature (Equation (7)).

This thermodynamic assessment is completed with a kinetic study of the Cr_2O_3 and Cl_2+O_2 interaction using TG isothermal tests and varying the chlorine content from 0% to 100% Cl_2. Three typical TGA curves at 50%, 80% and 100% Cl_2 are plotted in Figure 11a as evolution of %ML versus reaction time. More than 160 min were required to reach 75% of the Cr_2O_3 sample reacted in Cl_2 alone, while the reaction time is decreased to about 112 min when an equimolar (50% Cl_2+50% O_2) gas mixture was used and decreased again to about 85 min when the chlorine content in the Cl_2+O_2 was 80%. Data displayed in Figure 11b demonstrates that the initial reaction rate ($0.05 \leq \alpha \leq 0.40$) of the Cr_2O_3 chlorination with Cl_2+O_2 had a maximum value at 80% Cl_2 corresponding to Cl_2 to O_2 ratio equal to 4. Such a result matches well with the stoichiometric coefficients of the reaction described by Equation (14) resulting in chromyl chloride as final reaction product.

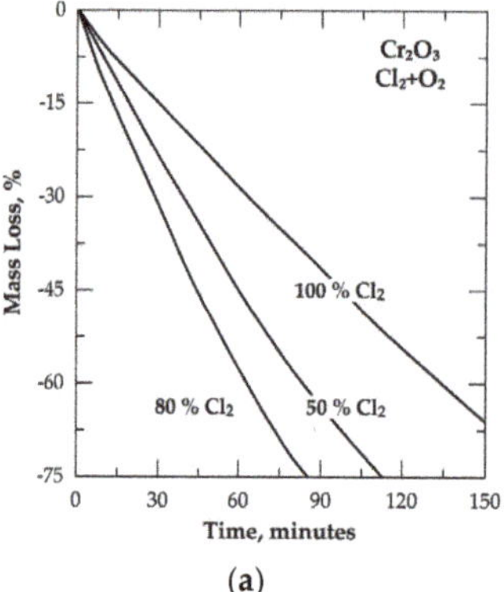

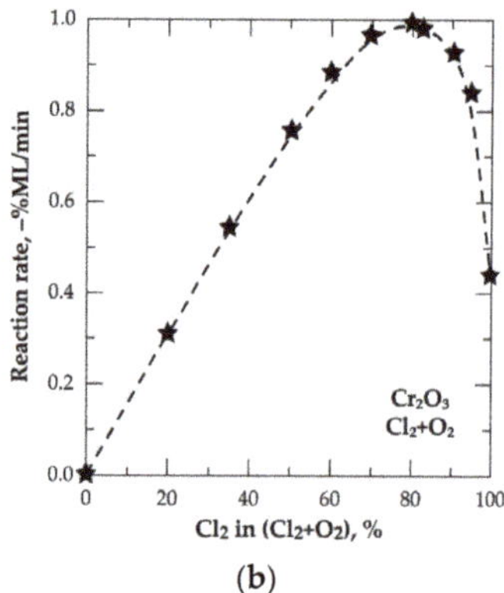

Figure 11. Treatment of Cr_2O_3 in Cl_2+O_2 at 800 °C: (**a**) Evolution of the sample mass loss versus time; (**b**) Dependency of the initial reaction rate on the chlorine content in the Cl_2+O_2 gas mixture.

According to this analysis, the LGChC is treated under a flowing gas (Cl_2+air) with total flow rate of 61 L/h containing 28 L/h of chlorine and 33 L/h air (i.e., 26 L/h N_2 and 7 L/h O_2) corresponding to Cl_2/O_2 molar ratio equal to 4. Complementary data for choosing this gas mixture composition to enhance the chromium oxide reaction rate and to lower the reaction of iron oxides with the chlorinating gas mixtures are found in previously published research reports [31–34].

Shown in Figure 12 is the %ML versus time for the isothermal treatment of the LGChC under the above-mentioned atmosphere between 950 and 1040 °C for reaction time up to 200 min. As in the previous cases (Figures 5 and 6), there is an abrupt change on the curve shape after 50% ML, it is more obvious at low temperature and reflected on the reaction progress. For instance, the first 50% ML was reached with a reaction time of 36 min at 950 °C, while more than 410 min are needed to reach 75% ML of the sample.

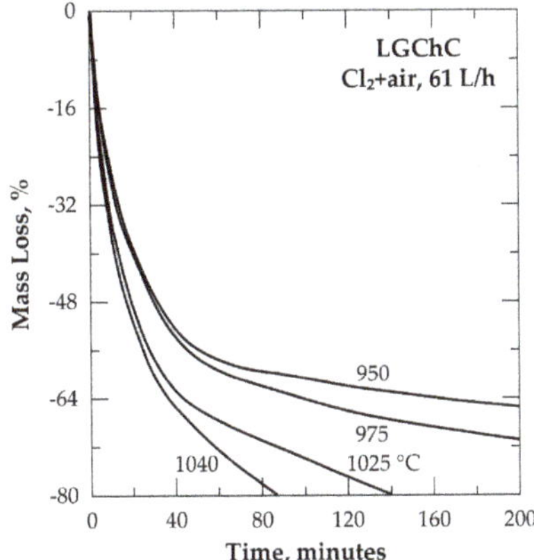

Figure 12. Mass change of the sample versus time for the treatment of LGChC in Cl$_2$+air.

For comparison, the acquired TG data for the reactions of LGChC and HGChC with chlorine in presence of oxygen are shown in Figure 13. The main observations are the high difference between %ML of HGChC and LGChC and the slope change of the %ML vs. time (ML rate), which is clearly more pronounced at 950 °C for both materials, although this evidence is also visible at 1025 °C. Combining the chemical and mineralogical composition of both concentrates with these isoconversion data lead to assign the kinetic changes to the substantial reaction of MgCr$_2$O$_4$ with chlorine for the conversion higher than 35% and 50 % for HGChC and LGChC, respectively. Both thermodynamic and kinetics reactivity may explain this particular behaviour of magnesiochromite in the Cl$_2$+air gaseous mixture.

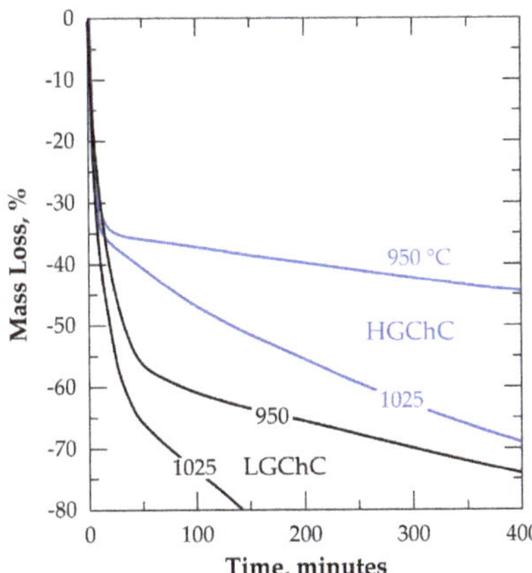

Figure 13. Comparison of the isothermal data for LGChC and HGChC treatment at 950 and 1025 °C under chlorine in presence of oxygen.

To have an idea about the evolution of the elemental and mineralogical composition of the treatment residue, the HGChC reacted at various α-values was examined by SEM-EDS (Figure 14). This method of analysis was preferred to XRD due to the presence of chromite crystalline isomorph phases.

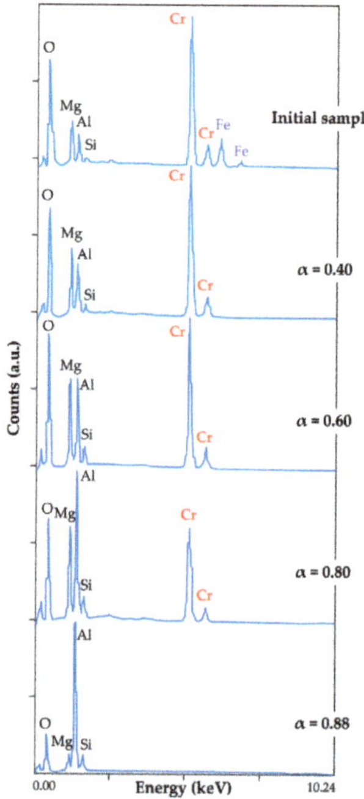

Figure 14. SEM-EDS results of residue from the oxychlorination of HGChC at different α-values.

Distinct characteristic peaks of Cr, Mg, Al, Fe, Si and O are present in the SEM-EDS spectrum of the initial sample, in good agreement with the HGChC elemental composition afore-mentioned in Figure 3. The product corresponding to the fraction reacted α = 0.40 does not contain iron. As iron bearing compounds of the chromite body are $FeO \cdot Fe_2O_3$ and $FeO \cdot Cr_2O_3$, the SEM-EDS analysis of the fraction reacted at α-0.40 is an indirect confirmation for the removal of magnetite and iron chromite from the HGChC. Spectra at α = 0.60 and α = 0.80 with their decreasing chromium peak intensity reflect the evolution of the composition, down to a chromium-free residue at α = 0.88. Accordingly, magnesio-chromite ($MgCr_2O_4$) had reacted at 0.35 < α < 0.88, while the true spinel (α = 0.88) appeared more refractory to chlorine in presence of oxygen. Reasoning by analogy, the reaction pathways of the LGChC with chlorine in presence of oxygen will be similar, with the conversion fraction agreeing with its mineralogical composition (15.1% Fe_3O_4, 37.6% $FeCr_2O_4$, 26.4% $MgCr_2O_4$, 20.8% $MgAl_2O_4$) displayed in Figure 4. The reaction rates are higher due to the fast reaction of 52.7% ($Fe_3O_4 + FeCr_2O_4$) with chlorine, hence exhibiting more reactive surface for the progress of the reaction.

The Arrhenius plot shown in Figure 15 for the reaction of LGChC with Cl_2+air at various α-values illustrates large changes in the apparent activation energy, starting from about 130 kJ/mol at beginning of the reaction, followed by an average E_a of about 80–85 kJ/mol at 0.15 ≤ α ≤ 0.50 and by a final strong increase up to 300 kJ/mol.

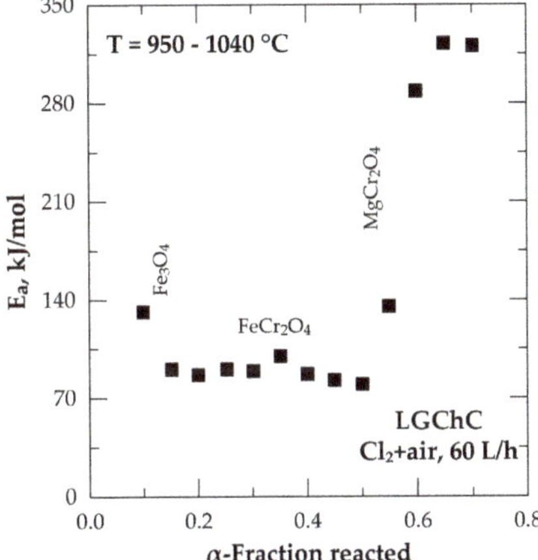

Figure 15. Plot of the apparent activation energy (E_a) versus fraction reacted for the treatment of LGChC with Cl_2+air at 950–1040 °C.

To help to interpret these peculiar changes, isothermal treatments under Cl_2+O_2 (Cl_2/O_2 = 4) of the main oxides (Fe_2O_3, Cr_2O_3 and MgO) of the chromite constituents (Fe_3O_4, $FeCr_2O_4$ and $MgCr_2O_4$) were performed up to 1025 °C. Note that ferrous oxide (FeO) is transformed into $FeCl_3$ and Fe_2O_3 under a chlorinating atmosphere [32]. The experimental data showed that the reactivity of these oxides towards Cl_2+O_2 is widely different. As an example, 90% of the Fe_2O_3 sample was reacted for 10 min at 950 °C. This reaction time, for reaching the same reaction extent, was extended to 60 min and 270 min for Cr_2O_3 and MgO, respectively at 950 °C. As shown in Figure 16a, this trend of the TGA measurements is still evident during the treatment of these oxides at 1000 °C. Based on the isothermal data recorded, the Arrhenius plots for the reactions of the above-mentioned oxides with Cl_2+O_2 are reported in Figure 16b. The E_a values are about 148, 46 and 214 kJ/mol for the Fe_2O_3, Cr_2O_3 and MgO reactions, respectively. In addition, the decreasing reaction rate ranking of simple oxides with Cl_2+O_2 follows the sequence represented by Equation (15).

$$Fe_2O_3 > Cr_2O_3 > MgO \tag{15}$$

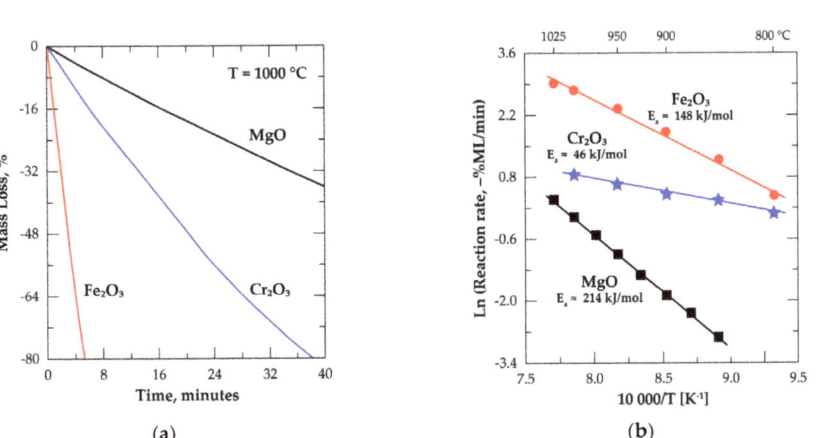

(a)

(b)

Figure 16. Treatment of Fe_2O_3, Cr_2O_3 and MgO in chlorine in presence of oxygen: (**a**) Evolution of the sample mass loss versus time at 1000 °C; (**b**) Arrhenius diagrams between 800 and 1025 °C.

Assuming that the chromite constituents (FeO·Fe$_2$O$_3$, FeO·Cr$_2$O$_3$, MgO·Cr$_2$O$_3$) are chlorinated only when both constituents (FeO and Fe$_2$O$_3$; FeO and Cr$_2$O$_3$; MgO and Cr$_2$O$_3$) are chlorinated and taking into account that the whole reaction rate of each chromite constituent is governed by the slowest reaction rate of its simple constituent, one may deduce that:

- the first compounds to be reacted are iron oxides of Fe$_3$O$_4$, the value of E$_a$ obtained for Fe$_2$O$_3$ reaction are close to that obtained for the beginning of the chromite reaction with Cl$_2$+O$_2$,
- the reaction of chromite with Cl$_2$+O$_2$, for $0.15 \leq \alpha \leq 0.50$, is controlled by the reaction rate of Cr$_2$O$_3$ (contained in FeCr$_2$O$_4$) being the slowest step of the overall reaction rate. The lower value of E$_a$ seems another argument for Cr$_2$O$_3$ impact on the overall reaction of chromite,
- the rest of the chromite reaction with Cl$_2$+O$_2$ ($\alpha > 0.50$) is affected by the reaction of MgO (MgCr$_2$O$_4$) which is characterized by a low reaction rate and a high E$_a$ value.

However, the energy of the chemical binding of the simple constituents in the chromite structure may affect the values of the inherent activation energy and the multistep reaction rates.

The comparison of the kinetic parameters for the reaction of chromite with chlorine and Cl$_2$+air showed a difference in the apparent activation energy values (Figures 10 and 15) essentially for α between 0.20 and 0.55. Two factors may explain that. First, from a thermodynamic point of view, the reaction of Cr$_2$O$_3$ with Cl$_2$ (Equation (7)) is less favourable than that with Cl$_2$+O$_2$ (Equation (14)); the E$_a$ values accordingly appear higher for the FeCr$_2$O$_4$ chlorination with Cl$_2$ alone. Second, the increasing and higher apparent E$_a$ with Cl$_2$ alone may reflect a MgCr$_2$O$_4$ reaction, characterized by a high E$_a$, starting before the FeCr$_2$O$_4$ reaction is finished.

In spite of these differences, the present study showed an atypical temperature impact on the chlorination of chromite due to the combination of different thermodynamic and kinetics aspects of chromite component reactions with chlorine in absence and/or presence of the oxygen.

As described in the previous sections, TGA tests were performed with small powder samples (40–50 mg) and high flowrates (e.g., 60 L/h Cl$_2$) of reactive gases to attenuate the reaction starvation impact and to enhance mass and heat transfers. To be closer to the practical chlorination process, tests using tenth grams of chromite concentrates were also performed in a horizontal setup described earlier [2] under Cl$_2$+air atmosphere from 700 to 1000 °C. The obtained data are depicted in Figure 17 as evolution of the chromium and iron content and chromium to iron ratio of the residues showing the function of the treatment temperature. These results show the preferential removal of iron from 700 °C; about 77% of iron and 18% of chromium were extracted during the treatment of HGChC at 900 °C for 2 h. The chromium and iron contents of the obtained residue at 900 °C were 35.6 and 3.2 wt%, respectively. As shown in Figure 17, the chromium to iron ratio increased rapidly from 3.9 at 700 °C to reach values as high as 11.1 at 900 °C.

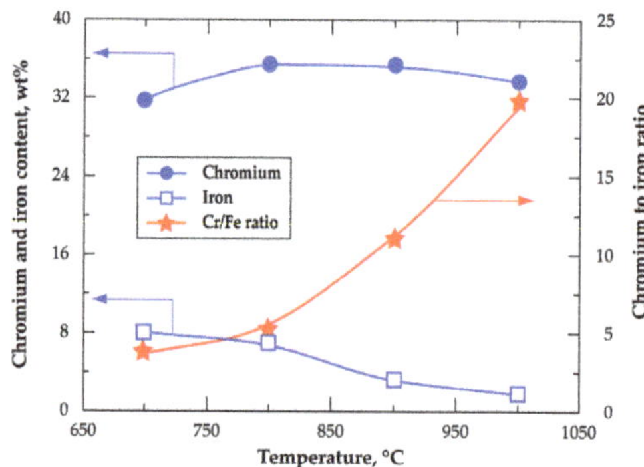

Figure 17. Evolution of chromium and iron content and chromium to iron ratio versus temperature during treatment of HGChC in a Cl$_2$+air gaseous mixture.

To gain understanding of the physical state of the chromium bearing phase synthetized during reaction of chromite with chlorine in presence of oxygen, a two-step cooling of the outlet gases was performed, first at room temperature and second at much lower temperature (−35° C).

The analyses of the solid condensate obtained at room temperature by SEM-EDS technique indicated the absence of chromium in the solid phase. However, a red liquid was isolated and collected (Figure 18) in a glassware vessel emerged in a refrigerated alcohol bath. This liquid produced reddish brown fumes in air and seems to correspond to chromyl chloride (CrO_2Cl_2) characteristics containing chromium at hexavalent state and it is characterized by a high vapor pressure at room temperature (Figure 5). The CrO_2Cl_2 synthesis was also reported in other research works [35–40] during thermal treatment of various chromium bearing materials.

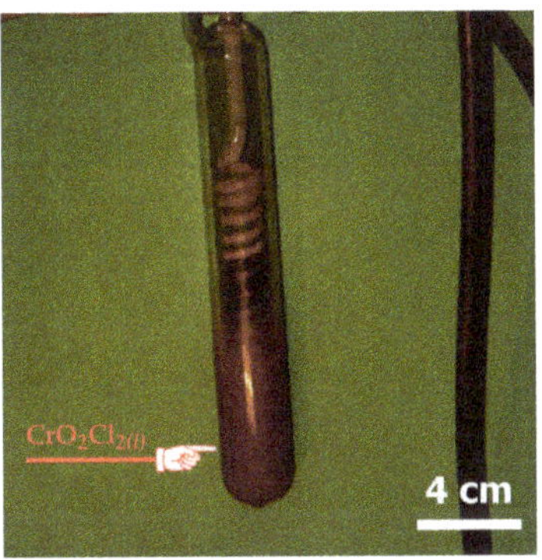

Figure 18. Visual image of the $CrO_2Cl_{2(l)}$ generated during the thermal treatment of chromite by Cl_2+air and collected in liquid state at −35 °C.

By analogy, the ability of chlorine to oxidize iron (II,III) into iron(VI) in high alkali medium by gas-solid and solid-solid reactions was also demonstrated in recent investigations [41–46].

This research work gave several insights for the evolution of the (Fe,Cr,Mg,Al)-O-Cl system, in the case of chromite, at different temperatures. Nonetheless, more in-depth and detailed studies are needed to complete the current knowledge in such a complex system.

4. Conclusions

Thermogravimetric analysis technique provides valuable information to fairly evaluate the constituent composition of complex materials such as chromite and to analyze its reactions with Cl_2 and Cl_2+O_2 gaseous mixtures.

The reactions of $(Fe,Mg)(Cr,Al,Fe)_2O_4$ constituents, for both chromite concentrates (low grade chromite concentrate-LGChC and high grade chromite concentrate-HGChC) with chlorine in isothermal conditions proceeded by gradual scheme starting by the reaction of iron oxides (Fe_3O_4) followed by interaction of iron chromite ($FeCr_2O_4$). Magnesio-chromite ($MgCr_2O_4$) appeared stable in Cl_2 and Cl_2+O_2 atmosphere at temperatures equal to or lower than 800 °C.

The overall reaction of LGChC with chlorine is affected differentially by temperature at 950–1040 °C, resulting in an apparent activation energy strongly dependent on the degree of conversion, e.g., increasing sharply from about 60 kJ/mol to 300 kJ/mol for fractions reacted of 0.15 and 0.75, respectively. Having a low reactivity, the $MgCr_2O_4$ compound required high temperature for the reaction to occur. Similar trends were observed for the reaction of chromite with chlorine in presence of oxygen although the values of the apparent activation energy are somewhat different.

Thermodynamic analysis of the envisaged reactions of the chromite constituents with Cl_2 and Cl_2+O_2 gave complementary elements for further clarifying this particular behavior of chromite in the chlorinating atmosphere.

The kinetics results of the simple oxides (Fe_2O_3, Cr_2O_3, MgO) reactions with Cl_2+O_2 in the interested temperature range was another insightful building block for better understanding the multistep process of chromite processing under chlorine in presence of oxygen.

Low temperatures and short times for the interaction chromite-chlorine favor the preferential removal of iron from the low-grade concentrate, giving a chromite with a chromium-to-iron ratio satisfactory for the ferrochrome production. The presence of oxygen in the system favors the synthesis of pure chromyl chloride.

Author Contributions: Conceptualization, N.K., E.A. and L.F.; Formal analysis, E.A., S.S. and F.D.; Investigation, N.K., E.A. and L.F.; Visualization, E.A., S.S. and F.D.; Resources, N.K., L.F. and F.P.; Writing—original draft, N.K., E.A., S.S., F.D. and F.P.; Writing—review and editing, N.K., E.A., L.F. and F.P. All authors have read and agreed to the published version of the manuscript.

Funding: Some data used in this paper were part of the PhD Thesis of Ndue Kanari. Another part of this development work has been supported by the French National Research Agency through the national program "Investissements d'avenir" with the reference ANR-10-LABX-21-01 / LABEX RESSOURCES21.

Conflicts of Interest: The authors declare no conflict of interest.

References

1. Critical Raw Materials—European Commission—Europa. Available online: https://ec.europa.eu/growth/sectors/raw-materials/specific-interest/critical_en (accessed on 29 March 2019).

2. Kanari, N. Extraction des Métaux de Valeur des Concentrés de Chalcopyrite et de Chromite par Chloruration. Ph.D. Thesis, Institut National Polytechnique de Lorraine, Vandœuvre-lès-Nancy, France, 7 November 1995; p. 209.

3. Kanari, N.; Gaballah, I.; Allain, E. A study of chromite carbochlorination kinetics. *Metall. Mat. Trans. B* **1999**, *30*, 577–587. [CrossRef]

4. Chromium Chapter—Handbook of Chemical Economics, Inorganic. Available online: https://pubs.usgs.gov/of/2001/0381/report.pdf (accessed on 27 July 2020).

5. Shen, S.-B.; Bergeron, M.; Richer-Laflèche, M. Effect of sodium chloride on the selective removal of iron from chromite by carbochlorination. *Int. J. Miner. Process.* **2009**, *91*, 74–80. [CrossRef]

6. Stainless Steel in Figures. 2019. Available online: https://www.worldstainless.org/Files/issf/non-image-files/PDF/ISSF_Stainless_Steel_in_Figures_2019_English_public_version.pdf (accessed on 28 July 2020).

7. Rifai, M.; Yuasa, M.; Miyamoto, H. Enhanced corrosion resistance of ultrafine-grained Fe-Cr alloys with subcritical Cr contents for passivity. *Metals* **2018**, *8*, 149. [CrossRef]

8. Chen, X.; Jiang, F.; Jiang, J.; Xu, P.; Tong, M.; Tang, Z. Precipitation, recrystallization, and evolution of annealing twins in a Cu-Cr-Zr alloy. *Metals* **2018**, *8*, 227. [CrossRef]

9. Tsarkov, A.A.; Zadorozhnyy, V.Y.; Solonin, A.N.; Louzguine-Luzgin, D.V. Effect of aluminum, iron and chromium alloying on the structure and mechanical properties of (Ti-Ni)-(Cu-Zr) crystalline/amorphous composite materials. *Metals* **2020**, *10*, 874. [CrossRef]

10. Di Schino, A.; Gaggiotti, M.; Testani, C. Heat Treatment Effect on Microstructure Evolution in a 7% Cr Steel for Forging. *Metals* **2020**, *10*, 808. [CrossRef]

11. Mishnev, R.; Dudova, N.; Kaibyshev, R.; Belyakov, A. On the fracture behavior of a creep resistant 10% Cr steel with high boron and low nitrogen contents at low temperatures. *Materials* **2020**, *13*, 3. [CrossRef] [PubMed]

12. Hao, M.; Sun, B.; Wang, H. High-temperature oxidation behavior of Fe–1Cr–0.2Si Steel. *Materials* **2020**, *13*, 509. [CrossRef]

13. Daniel, J.; Grossman, J.; Houdková, Š.; Bystrianský, M. Impact wear of the protective Cr_3C_2-based HVOF-sprayed coatings. *Materials* **2020**, *13*, 2132. [CrossRef]

14. Nafziger, R.H. A review of the deposits and beneficiation of the low-grade chromite. *J. S. Afr. Inst. Min. Metall.* **1982**, *8*, 205–225.

15. Zhao, Q.; Liu, C.J.; Zhang, B.; Jiang, M.; Qi, J.; Saxén, H.; Zevenhoven, R. Study on extraction of iron from chromite. *Steel Res. Int.* **2015**, *86*, 1541–1547. [CrossRef]

16. Hussein, M.K.; El-Barawi, K. Study of the chlorination and beneficiation of Egyptian chromite ores. *Trans.-Inst. Min. Metall. Sect. C* **1971**, *80*, C7–C11.

17. Hussein, M.K.; Winterhager, H.; Kammel, R.; El-Barawi, K. Chlorination behaviour of the main oxide components chromite ores. *Trans.-Inst. Min. Metall. Sect. C* **1974**, *83*, C154–C160.

18. Robinson, M.; Crosby, A.D. Production of Metal Chlorides. Europe Patent Application EP, No 0 096 241, 21 December 1983; 25p.

19. Martirosyan, V.A. Thermodynamics of the chlorination of individual oxides in chromium ores with hydrogen chloride. *Arm. Khim. Zh.* **1978**, *31*, 93–99.

20. Martirosyan, V.A. Thermodynamics of the chlorination of complex oxides in chromium ores with hydrogen chloride. *Arm. Khim. Zh.* **1978**, *31*, 100–106.

21. Vil'nyanskii, Y.E.; Martirossyan, V.A. Kinetics of selective chlorination of chromite ores by hydrogen chloride. *Arm. Khim. Zh.* **1973**, *26*, 881–888.

22. Kanari, N.; Gaballah, I.; Allain, E. Use of chlorination for chromite upgrading. *Thermochim. Acta* **2000**, *351*, 109–117. [CrossRef]

23. Kanari, N.; Gaballah, I.; Allain, E. Kinetics of oxychlorination of chromite Part, I. Effect of temperature. *Thermochim. Acta* **2001**, *371*, 143–154. [CrossRef]

24. Kanari, N.; Gaballah, I.; Allain, E. Kinetics of oxychlorination of chromite part II. Effect of reactive gases. *Thermochim. Acta* **2001**, *371*, 75–86. [CrossRef]

25. Allain, E. Recyclage des Pentoxydes de Tantale et Niobium Contenus dans les Scories de Four à étain par un Traitement Hydro-Pyrométallurgique. Ph.D. Thesis, Université de Nancy I, Nancy, France, June 1993; p. 136.

26. Kanari, N.; Allain, E.; Shallari, S.; Diot, F.; Diliberto, S.; Patisson, F.; Yvon, J. Thermochemical route for extraction and recycling of critical, strategic and high value elements from by-products and end-of-life materials, Part I: Treatment of a copper by-product in air atmosphere. *Materials* **2019**, *12*, 1625. [CrossRef]

27. Roine, A. *Outokumpu HSC Chemistry for Windows, Version 3.0*; Outokumpu Research: Pori, Finland, 1997.

28. Anonymous. *Handbook of Chemistry and Physics*, 66th ed.; Weast, R.C., Astle, M.J., Beyer, W.H., Eds.; CRC Press: Boca Raton, FL, USA, 1986; pp. D193–D194.

29. Pascal, P. *Nouveau Traité de Chimie Minérale*; Tome XIV, Masson et Compagnie: Paris, France, 1959; p. 135.

30. Fedunik-Hofman, L.; Bayon, A.; Donne, S.W. Kinetics of solid-gas reactions and their application to carbonate looping systems. *Energies* **2019**, *12*, 2981. [CrossRef]

31. Gaballah, I.; Ivanaj, S.; Kanari, N. Kinetics of chlorination and oxychlorination of chromium (III) oxide. *Metall. Mat. Trans. A* **1998**, *29*, 1299–1308. [CrossRef]

32. Kanari, N.; Allain, E.; Gaballah, I. Reactions of wüstite and hematite with different chlorinating agents. *Thermochim. Acta* **1999**, *335*, 79–86. [CrossRef]

33. Kanari, N.; Mishra, D.; Filippov, L.; Diot, F.; Mochón, J.; Allain, E. Kinetics of hematite chlorination with Cl_2 and Cl_2+O_2: Part, I. Chlorination with Cl_2. *Thermochim. Acta* **2010**, *497*, 52–59. [CrossRef]

34. Kanari, N.; Mishra, D.; Filippov, L.; Diot, F.; Mochón, J.; Allain, E. Kinetics of hematite chlorination with Cl_2 and Cl_2+O_2. Part II. Chlorination with Cl_2+O_2. *Thermochim. Acta* **2010**, *506*, 34–40. [CrossRef]

35. Saeki, Y.; Matsuzaki, R.; Morita, H. Chlorination of chromic oxide. *Kogyo Kagaku Zasshi* **1971**, *74*, 344–348. [CrossRef]

36. Sano, N.; Belton, G.R. The thermodynamics of volatilization of chromic oxide: Part II. The species CrO_2Cl_2. *Met. Trans.* **1974**, *5*, 2151–2154. [CrossRef]

37. Kuffa, T.; Ponevsky, N.; Skrobian, M. The chlorination kinetics of the chromic oxide. *Thermochim. Acta* **1985**, *92*, 201–204. [CrossRef]

38. Jacobson, N.S.; McNallan, M.J.; Lee, Y.Y. Mass spectrometric observations of metal oxychlorides produced by oxidation-chlorination reactions. *Metall. Trans. A* **1989**, *20*, 1566–1568. [CrossRef]

39. Chen, J.C.; Wey, M.Y.; Chiang, B.C.; Hsieh, S.M. The simulation of hexavalent chromium formation under various incineration conditions. *Chemosphere* **1998**, *36*, 1553–1564. [CrossRef]

40. Amrute, A.P.; Mondelli, C.; Pérez-Ramírez, J. Kinetic aspects and deactivation behaviour of chromia-based catalysts in hydrogen chloride oxidation. *Catal. Sci. Technol.* **2012**, *2*, 2057–2065. [CrossRef]

41. Kanari, N.; Evrard, O.; Neveux, N.; Ninane, L. Recycling ferrous sulfate via super-oxidant synthesis. *JOM* **2001**, *53*, 32–33. [CrossRef]

42. Kanari, N.; Ostrosi, E.; Ninane, L.; Neveux, N.; Evrard, O. Synthesizing alkali ferrates using a waste as a raw material. *JOM* **2005**, *57*, 39–42. [CrossRef]

43. Ninane, L.; Kanari, N.; Criado, C.; Jeannot, C.; Evrard, O.; Neveux, N. New processes for alkali ferrate synthesis. In *Ferrates: Synthesis, Properties and Applications in Water and Wastewater Treatment*; Sharma, V.K., Ed.; American Chemical Society: Washington, DC, USA, 2008; pp. 102–111.

44. Kanari, N.; Filippov, L.; Diot, F.; Mochón, J.; Ruiz-Bustinza, I.; Allain, E.; Yvon, J. Synthesis of potassium ferrate using residual ferrous sulfate as iron bearing material. *J. Phys. Conf. Ser.* **2013**, *416*, 012013. [CrossRef]

45. Kanari, N.; Filippova, I.; Diot, F.; Mochón, J.; Ruiz-Bustinza, I.; Allain, E.; Yvon, J. Utilization of a waste from titanium oxide industry for the synthesis of sodium ferrate by gas-solid reactions. *Thermochim. Acta* **2014**, *575*, 219–225. [CrossRef]

46. Kanari, N.; Ostrosi, E.; Diliberto, C.; Filippova, I.; Shallari, S.; Allain, E.; Diot, F.; Patisson, F.; Yvon, J. Green process for industrial waste transformation into super-oxidizing materials named alkali metal ferrates (VI). *Materials* **2019**, *12*, 1977. [CrossRef] [PubMed]

© 2020 by the authors. Licensee MDPI, Basel, Switzerland. This article is an open access article distributed under the terms and conditions of the Creative Commons Attribution (CC BY) license (http://creativecommons.org/licenses/by/4.0/).

 *materials*

Article

High-Temperature Mechanical Behaviors of SiO$_2$-Based Ceramic Core for Directional Solidification of Turbine Blades

Jiangwei Zhong and Qingyan Xu *

Key Laboratory for Advanced Materials Processing Technology (MOE), School of Materials Science and Engineering, Tsinghua University, Beijing 100084, China; zhongjw16@mails.tsinghua.edu.cn
* Correspondence: scjxqy@tsinghua.edu.cn

Received: 13 August 2020; Accepted: 10 October 2020; Published: 14 October 2020

Abstract: The high-temperature mechanical behaviors of SiO$_2$-based ceramic cores for the directional solidification of turbine hollow blades were investigated. Isothermal uniaxial compression tests of ceramic core samples were conducted on a Gleeble-1500D mechanical simulator with an innovative auxiliary thermal system. The stress–strain results and macro- and micro- structures of SiO$_2$-based ceramic cores were investigated experimentally. The microstructures were characterized by the scanning electron microscope (SEM). Based on the experimental data, a nonlinear constitutive model for high temperature compressive damage was established. The statistical results of Weibull moduli show that the stability of hot deformation increases with the increase of temperature. The fracture type of the SiO$_2$-based core samples is brittle fracture, but when the temperature exceeds 1400 °C, the mechanical behavior exhibits thermo-viscoelastic and viscoplastic property. Under high-temperature (>1400 °C) and stress conditions, the strength of the ceramic core is weakened owing to the viscous slip of SiO$_2$, which is initially melted at the temperature of 1400 °C. The comparison results between the predictions of nonlinear model and experimental values indicate that the model is applicable.

Keywords: superalloy; ceramic core; high temperature; mechanical behavior; auxiliary thermal system

---

1. Introduction

In response to the increasing worldwide need for reliable, low-cost, and environmentally compatible generation of energy, the new generation of H-class gas turbines (GT) is developed [1,2]. Ni-based single-crystal (SX) superalloy turbine blades, which are the key hot-end assemblies of the gas turbine engines, can be produced by using the directional solidification (DS) technology [3]. The complex inner cavity formed by the ceramic core provides the possibility for the development of the hollow blade cooling technology. Nowadays, due to the complex thermal stress–strain interactions during DS, the size of blades appears imprecise, and the ceramic core even appears cracked. As a result, the performance of the SiO$_2$-based ceramic core directly affects the dimensional accuracy of the SX hollow turbine blade. Therefore, the high-temperature mechanical properties of the SiO$_2$-based ceramic core are crucial for the preparation of SX hollow turbine blade.

There are some studies focused on the high-temperature mechanical behaviors of SiO$_2$-based ceramic cores [4]. Xu et al. [5] investigated the flexural strength of silica-based ceramic cores doped with different silica nanopowders at 1540 °C. The results showed that large quantities of cristobalite were crystallized at 1540 °C, which might provide enhanced mechanical property in the casting. Niu et al. [6] found that the ceramic cores with 3 wt% mullite fibers showed excellent properties, such as flexural strength being 22.3 MPa at 1550 °C, owing to fiber reinforcing. At the same time, there are two opposite conclusions about the high-temperature properties of SiO$_2$-based ceramic core. For instance, Kazemi et al. [7] found that the increase of zircon content could result in the decrease of cristobalite formed in situ, owing to cristobalite crystallized on the surface of fused silica particles during heat treatment. This result is contrary to Wang and Hon's report [8], but is in good agreement with the result of Wilson et al. [9]. On the other hand, there are many examples in the literature exploring the static and dynamic hot deformation behaviors of other types of ceramic materials in-depth, such as ZrB_2 [10], SiC [11], Al_2O_3 [12], and ceramic composite material [13]. The methods presented in these articles can be applied to the study of SiO_2-based ceramic core.

During the directional solidification process, the ceramic cores will be subjected to mechanical loading at high temperature for a long time. In order to prevent the core fracture, it is very necessary to investigate the high temperature behavior of ceramic core. In this study, an auxiliary thermal system is employed to carry out the thermal compression tests of ceramic core. Constitutive modeling and various characterization methods are used to understand the high-temperature mechanical property of SiO_2-based ceramic core.

2. Experimental Procedure

2.1. Experimental Methods and Design

The characteristics of fused silica and zircon used as raw materials are illustrated in Table 1. According to the formulation used in actual production, the composition of the samples was 60 wt% fused silica and 40 wt% zircon. Porous silica-based ceramic cores were prepared by using ceramic injection molding. After a series of procedures, such as mixing, ball milling, adding adhesive, and drying, the green bodies were obtained. The sintered samples were subsequently subjected to heat treatment at 1500 °C for 30 min and then were removed from the furnace at 1000 °C, to the atmosphere at 25 °C, for simulating the realistic rapid cooling process during the directional solidification from the heating zone into the cooling zone.

Table 1. Characteristics of the used fused silica and zircon as raw materials.

Powder (%)	SiO_2	ZrO_2	Al_2O_3	K_2O	CaO	TiO_2	Powder Density (g/cm^3)	Open Porosity (%)	Source
Fused Silica	99.99	-	0.002	-	0.004	0.001	1.99	0.8	HongDa
Zircon	33.21	62.5	0.82	0.90	0.55	1.94	4.54	1.1	XinTai

Thermal process of ceramic shell/core during directional solidification is shown in Figure 1a. Since the tendency of core breakage is mainly concentrated at temperatures above 800 °C, the hot compression temperatures of ceramic cores were set at 700 °C (ST700), 1100 °C (ST1100), and 1400 °C (ST1400). The sample ST25 was tested at 25 °C. The average size of the ceramic cores is 14.77 mm in diameter and 15.25 mm in length (φ14.77 mm × 15.25 mm). The stain rate of 0.001 s^{-1} was chosen. After taking into consideration the inhomogeneity of the ceramic core, we tested three parallel samples for each deformation temperature. The whole process can be represented by Figure 1 and Table 2.

2.2. High-Temperature Experimental System for Mechanical Behaviors

The Gleeble system has been widely employed in the research of material constitutive model [14,15]. It mainly includes three parts: heating system, mechanical system, and computer control system. The heating system forms a current loop with a loaded metal sample (as a resistor) to heat the metal sample. The heating rate and heating temperature are varied by controlling the current in the sample. Therefore, the Gleeble system is generally unable to measure the high-temperature mechanical properties of non-conductive materials, such as ceramic materials. In order to realize the high-temperature mechanical measurement of non-conductive materials on the Gleeble simulator, we designed and developed an auxiliary thermal device that could expand the measurement material range of the Gleeble simulator [16]. The schematic diagram of the auxiliary thermal system of the Gleeble testing is shown in Figure 2. The measured temperature can reach 1600 °C. The temperature control accuracy is ±4 °C.

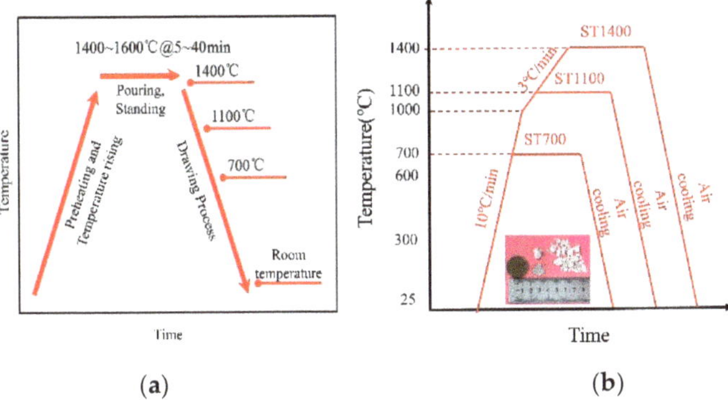

(**a**) (**b**)

Figure 1. (**a**) Thermal process of directional solidification; and (**b**) the schematic diagram illustrating the compression process of SiO$_2$-based ceramic cores.

Table 2. Heat treatments of ceramic cores to simulate the directional solidification process and test conditions.

Ceramic Core Samples	Sintering	Heat Treatment	Test Temperature
ST25	1000 °C @ 60 min	-	25 °C
ST700	1000 °C @ 60 min	1500 °C @ 30 min	700 °C
ST1100	1000 °C @ 60 min	1500 °C @ 30 min	1100 °C
ST1400	1000 °C @ 60 min	1500 °C @ 30 min	1400 °C

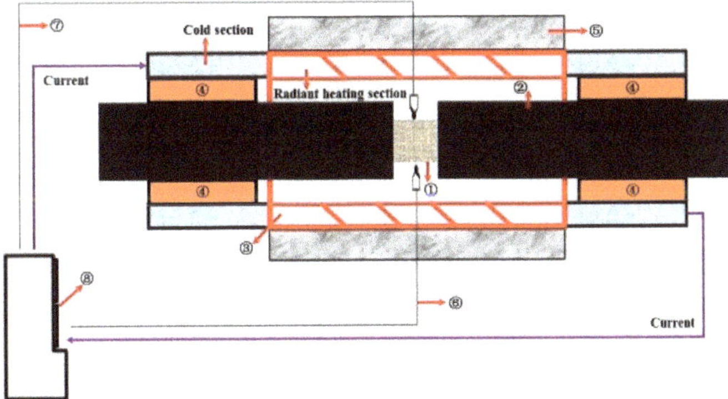

Figure 2. Schematic diagram of the auxiliary thermal system (① ceramic sample, ② compression bars, ③ silicon carbide screw tube, ④ corundum tube, ⑤ insulation fiber box, ⑥ S-type thermocouple for temperature control, ⑦ temperature S-type thermocouple for temperature calibration, and ⑧ temperature control cabinet).

3. Results and Discussion

3.1. High-Temperature Mechanical Properties

Figure 3 shows the stress–strain curves variation ranges of the samples ST25, ST700, ST1100, and ST1400, from which can be found ST25, ST700, and ST1100 are all brittle fractures, while ST1400 shows thermo-viscoelastic and viscoplastic property. The average compressive strengths of ST25, ST700, and ST1100 are 51.83, 55.82, and 80.46 MPa, respectively, as shown in Figure 4. It is worth noting that the compressive strength of ST1100 is higher than that of ST25 and ST700. This is mainly due to the conversion of α-cristobalite to β-cristobalite. The densities of α-cristobalite and β-cristobalite are about 2.32 and 2.22 g/cm^3, respectively, and the expansion of the β-cristobalite volume makes the sample more compact [17]. More α-cristobalite is transformed into β-cristobalite at 1100 °C, and the strength of β-cristobalite is stronger than that of α-cristobalite [18]. At the same time, Zener pinning would be more pronounced at higher temperatures [9].

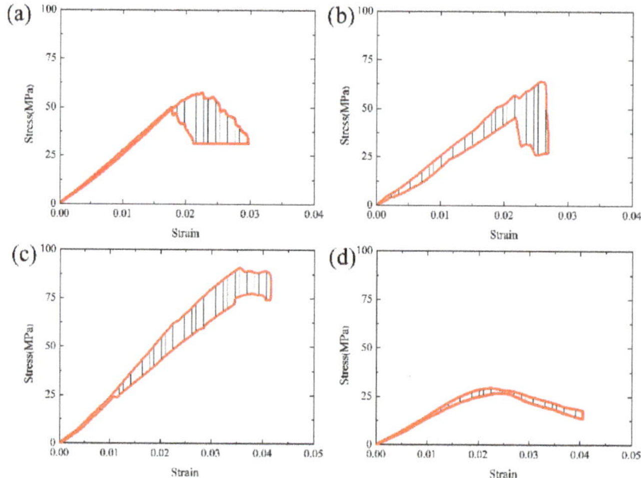

Figure 3. Stress–strain experimental data zone of isothermal uniaxial compression tests: (**a**) ST25, (**b**) ST700, (**c**) ST1100, and (**d**) ST1400.

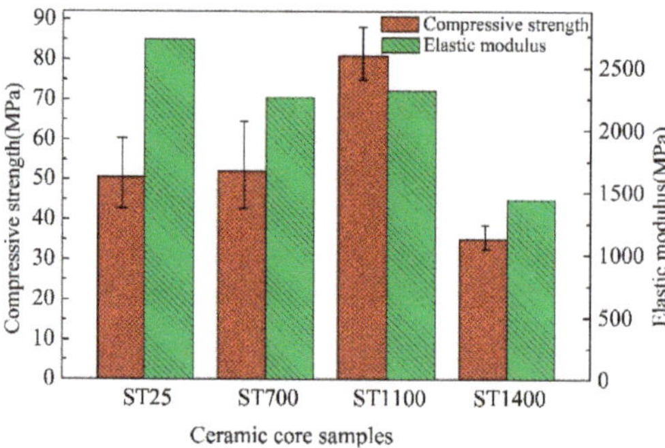

Figure 4. Compressive strength and elastic modulus of samples (ST25, ST700, ST1100, and ST1400).

In the elastic stage, the stress–strain curve of sample ST25 has a good linear-elastic regime. With the increase of temperature, the stress–strain curves of ST700 and ST1100 have a large amplitude and demonstrate a certain degree of dispersion. The elastic moduli of ST25, ST700, ST1100, and ST1400 are 2726, 2259, 2316, and 1442 MPa, respectively. The elastic modulus of the SiO$_2$-based ceramic core shows a decreasing trend when the temperature is increased. There is a small change in the range of 25~1100 °C, while the elastic modulus decreases rapidly at the range of 1100~1400 °C. The stress–strain curve of ST1400 at the viscoplastic stage is narrow, indicating that the high-temperature experimental result achieves high repeatability and reproducibility. However, the overall high-temperature mechanical property of the sample ST1400 decreases significantly.

On the other hand, it can be found from Figure 4 that the dispersion properties of compressive strengths vary greatly with the increase of temperature. The samples ST25 and ST700 demonstrate large dispersion, while samples ST1100 and ST1400, especially sample ST1400, show less dispersion. To quantify this result, a universal empirical model called Weibull approach is introduced.

The three-parameter Weibull distribution function can be simplified to a two-parameter Weibull distribution function without affecting its accuracy [10,19]:

$$P(x) = 1 - \exp\left[-(\sigma/\sigma_0)^m\right] \tag{1}$$

where $P(x)$ is the failure probability, σ_0 is shape factor, and m is Weibull modulus. Generally speaking, the larger m indicates the material is more uniform and less dispersion.

From Figure 5, it can be found the Weibull moduli of ST25, ST700, ST1100, and ST1400 are 12.24, 10.87, 18.65, and 38.39, respectively. Obviously, ST1400 demonstrates the largest modulus, indicating that the hot deformation of sample at 1400 °C tends to exhibit certain stable and repeatable characteristic. At the same time, as the moduli of ST700 and ST25 are relatively small, it can be concluded that the deformation at 700 and 25 °C will be quite unstable. In fact, these results further confirm the difference of stress–strain curves in Figure 3 from the angle of data.

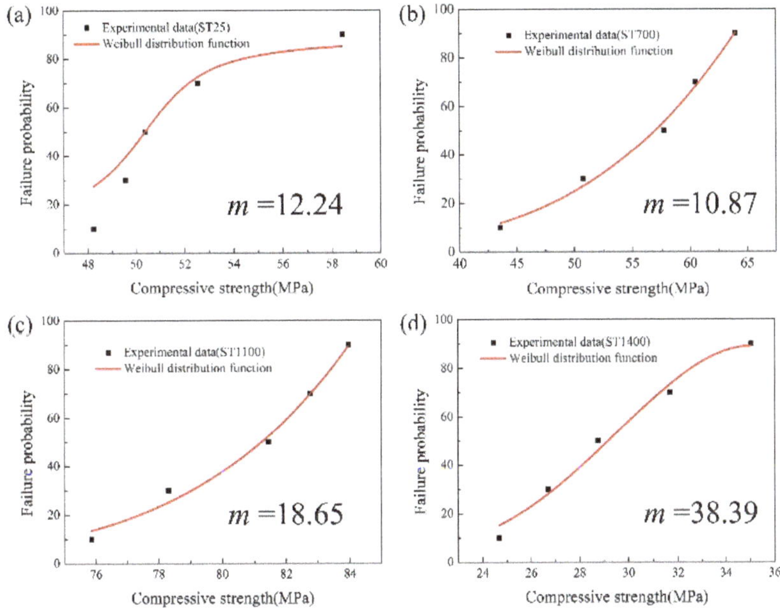

Figure 5. Weibull distribution of high-temperature compressive strengths for SiO_2 ceramic cores: (**a**) ST25, (**b**) ST700, (**c**) ST1100, and (**d**) ST1400.

3.2. Microstructures Evolution

As is well-known, the mechanical properties are always associated with microstructures evolution. Figure 6 exhibits the macrostructural investigations of the samples (ST25, ST700, ST1100, and ST1400) fracture surfaces. The difference in the fracture patterns of ST25, ST700, ST1100, and ST1400 can be clearly distinguished (Figure 6a–d). ST25, ST700, and ST1100 are crushed brittle fractures. As the temperature increases, the average size of residual pieces increases. When the strain of ST1400 reaches 0.04, the sample can maintain a substantially complete shape, and only minor fragments occur on the cylindrical surface. At the same time, a sliding plane of approximately 45° with the bottom surface of the cylindrical sample can be clearly seen. It is this kind of viscous slip that causes the stress plateau of ST1400 after the strain is greater than 0.02 at 1400 °C. The main reason of this sticky slip can be explained in the microstructure morphology.

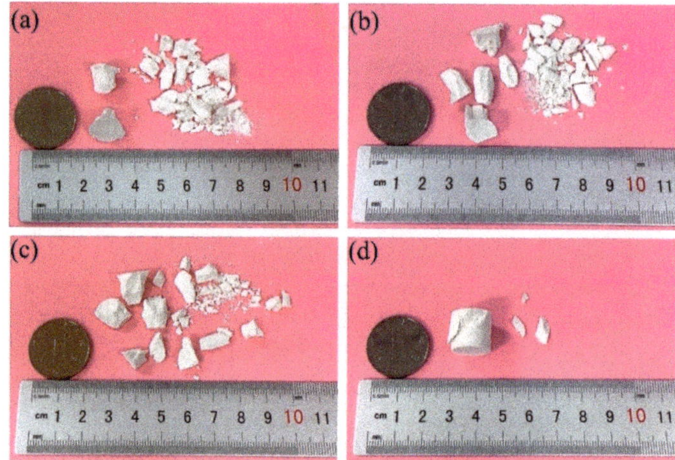

Figure 6. Macrostructural investigation of the sample fracture: (**a**) ST25, (**b**) ST700, (**c**) ST1100, and (**d**) ST1400.

Figure 7 shows the XRD patterns of various samples after hot compression at different temperatures. It can be found that all samples are composed of zircon and α-cristobalite. With the increase of deformation temperature, the peak intensities of two phases have a little change. In order to clarify the phase distribution in the microstructure, EDS point elemental analysis was employed and the results are shown in Figure 8. As shown in Figure 8, the point 1 with gray-black color is SiO_2 and the point 2 with bright white color is $ZrSiO_4$ in the image of backscattered electron (BSE).

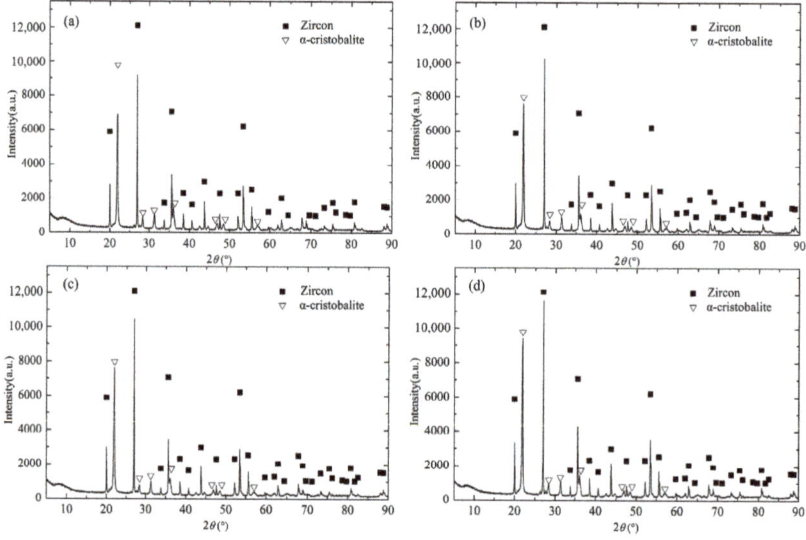

Figure 7. XRD patterns of ceramic samples: (**a**) ST25, (**b**) ST700, (**c**) ST1100, and (**d**) ST1400.

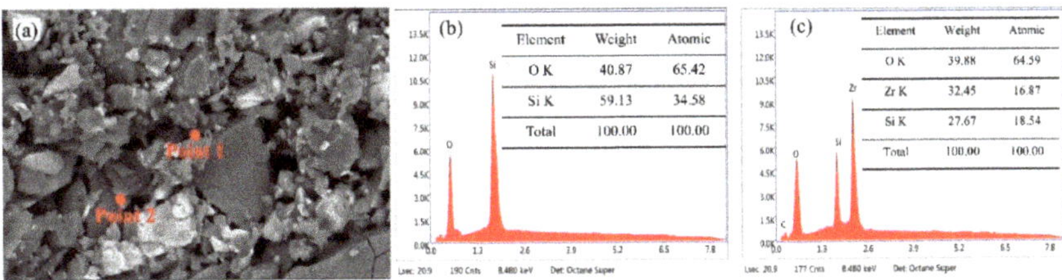

Figure 8. EDS analysis of ST1100: (**a**) the image of BSE, (**b**) EDS result of point 1, and (**c**) EDS result of point 2.

Figure 9a,c,e,g shows the images of secondary electrons (SE), and Figure 9b,d,f,h shows the images of the BSE. In the BSE images, the major component of the white-gray phase is $ZrSiO_4$, and the major component of the black-gray phase is SiO_2. From the micro-topography of ST25, ST700, and ST1100, it can be seen that the SiO_2 particles mainly undergo cleavage fracture, the $ZrSiO_4$ particles mainly undergo dimple fracture. Most penetrated cracks are distributed in larger SiO_2 particles. However, there are almost no cleavage fractures in SiO_2 particles of the ST1400, and there are little dimple-like $ZrSiO_4$ sections. The surfaces of the small SiO_2 particles have a smooth curved shape. The occurrence of material fracture usually has great uncertainly, which is also the reason for the divergence of the stress–strain curves. As mentioned before, the Weibull modulus of ST1400 is much larger than that of ST25, ST700, and ST1100, meaning that the deformation of ST1400 is more stable. Therefore, the microstructure observation results of samples above are relatively consistent with the stress–strain curves.

In the cross-section of ST25, ST700, and ST1100, the surfaces of the large SiO_2 particles are clean, and almost no adhesion of fine SiO_2 particles is observed. However, in addition to penetrating cracks on the surface of the large SiO_2 particles in ST1400, a large number of smooth and fine SiO_2 particles are attached to the surface (Figure 9h). It is generally believed that the melting temperature of β-cristobalite is $1720 \pm 10\ °C$ [17,20]. However, some studies have shown that, when the temperature reaches 1400~1450 °C, it will slowly melt on the surface of the SiO_2, and the presence of other elements or impurities may reduce the β-cristobalite transformation temperature [21,22]. Therefore, in the high-temperature environment of 1400 °C, the main reason for the viscous slip of the SiO_2-based ceramic core samples is that the surfaces of the fine SiO_2 particles are initially melted, which plays a role in lubrication between large particles. The SiO_2, which is initially melted at the temperature of 1400 °C, adheres to the surface of the large SiO_2 particle. When the temperature drops further to the room temperature, it combines with the large particles, to form a unitary body.

3.3. Nonlinear Constitutive Models for High-Temperature Compressive Damage

It can be seen from Figures 4 and 5 that the compressive strength and modulus are basically negatively correlated with temperature. The macro-effect of temperature on the properties of ceramic core includes two aspects. On the one hand, the intermolecular forces decrease with increasing temperature. On the other hand, the change of structure caused by the variation of temperature will greatly affect the properties of the material, such as thermal mismatch. Therefore, the thermal damage $D(T)$ is employed to describe the temperature effect on the property [23]:

$$D(T) = 1 - E_T/E_0 \tag{2}$$

where E_0 is elastic modulus at room temperature, and E_T is the elastic modulus at T. The elastic modulus denoted by thermal damage is expressed as follows:

$$E_T = [1 - D(T)]E_0 \tag{3}$$

According the analysis of the experimental results at different temperatures, the thermal damage value, *D(T)*, at each temperature point is calculated, as shown in Figure 10. Through data fitting, the expression of thermal damage with temperature variation can be written as follows:

$$D(T) = -0.0328 + 0.00125T - 2.136 \times 10^{-6}T^2 + 1.068 \times 10^{-9}T^3 \tag{4}$$

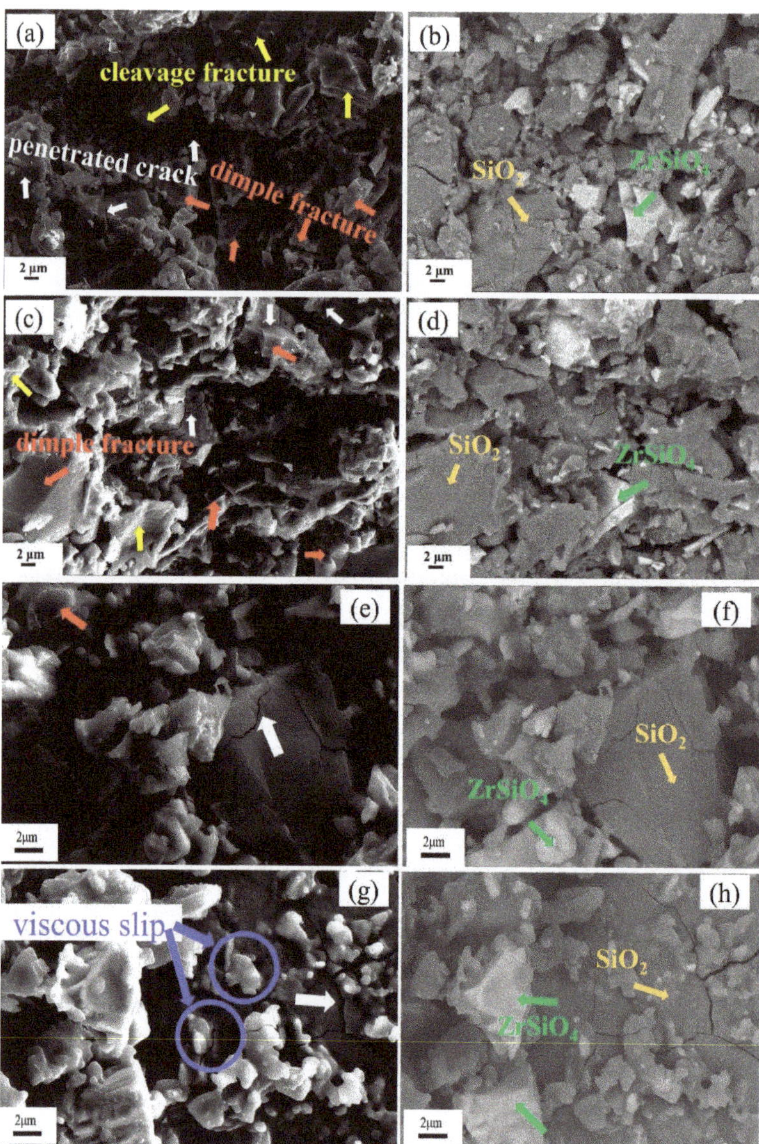

Figure 9. SEM images of the sample fracture surface (red arrows = dimple fracture; yellow arrows = cleavage fracture; blue arrows = high temperature viscous slip; white-gray = ZrSiO₄, and black-gray = SiO₂). SE: (**a**) ST25, (**c**) ST700, (**e**) ST1100, and (**g**) ST1400. BSE: (**b**) ST25, (**d**) ST700, (**f**) ST1100, and (**h**) ST1400.

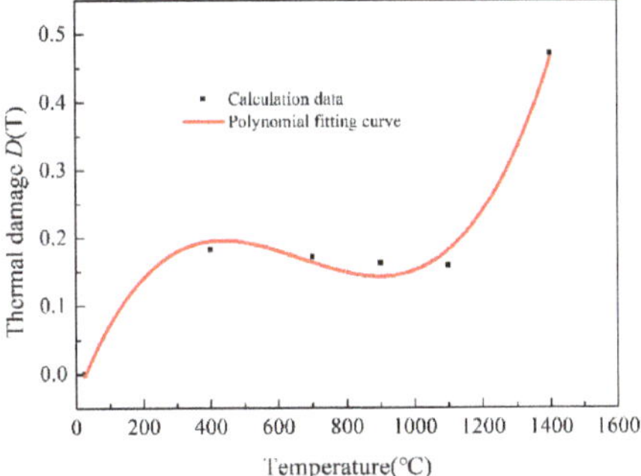

Figure 10. Thermal damage values of ceramic cores under different temperatures.

According to the author's previous research [24], the continuous damage constitutive model based on Weibull distribution method, at room temperature, is summarized as follows:

$$\sigma_1 = E\varepsilon_1 \exp\left[-\left(\frac{\varepsilon_1}{\varepsilon_0}\right)^m\right]. \tag{5}$$

where ε_1 is the axial strain, and ε_0 is a constant. In order to obtain the nonlinear constitutive model for high-temperature compressive damage, the E in Equation (5) can be substituted by E_T, and the formula can be rewritten as follows:

$$\sigma_1 = [1 - D(T)]E_0\varepsilon_1 \exp\left[-\left(\frac{\varepsilon_1}{\varepsilon_0}\right)^m\right] \tag{6}$$

The experiment results of typical compression stress–strain of SiO_2-based ceramic core and the simulation results based on thermo-visco damage model are presented in Figure 11.

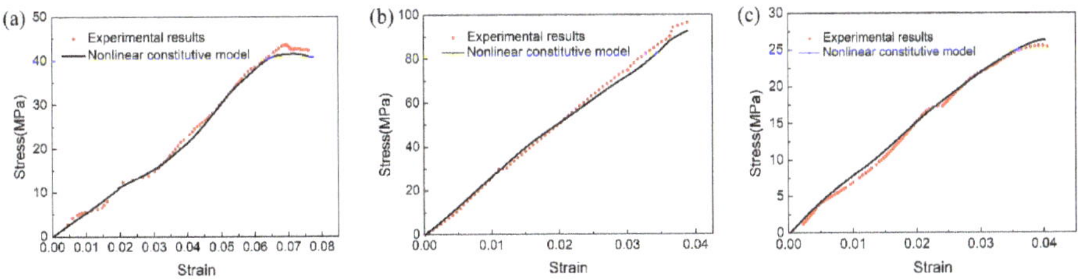

Figure 11. The comparison between the experimental results and nonlinear constitutive model prediction: (**a**) ST700, (**b**) ST1100, and (**c**) ST1400.

From Figure 11, it can be found that the nonlinear constitutive model has a good generalization property. In other words, this model could reflect the uniaxial compression behaviors of ceramic cores deformed at various temperatures.

4. Conclusions

(1) In the temperature range from 25 to 1400 °C, the elastic moduli of the SiO_2-based ceramic cores range from 1442 to 2726 MPa at the elastic stages. The statistical results of Weibull moduli show that the stability of deformation increases with the increase of temperature.

(2) The SiO_2-based ceramic core samples are all brittle fractures, while, when the temperature exceeds 1400 °C, the mechanical behaviors of the samples are characterized by thermo-viscoelastic

and viscoplastic properties, which mainly can be ascribed to the initial surface melting of SiO_2 fine particles.

(3) Nonlinear constitutive model for high-temperature compressive damage is established to predict the hot deformation of ceramic core. The comparison results between the nonlinear model predictions and experimental values indicate that the model is applicable.

Author Contributions: Conceptualization, J.Z.; data curation, J.Z.; formal analysis, J.Z.; funding acquisition, Q.X.; validation, J.Z.; writing—original draft, J.Z.; writing—review and editing, Q.X. All authors have read and agreed to the published version of the manuscript.

Funding: This work was supported by the National Science and Technology Major Project (2017-VI-0003-0073 and 2017-VII-0008-0101) and the Project to Strengthen Industrial Development at the Grass-Roots Level (TC160A310/18).

Conflicts of Interest: The authors declare no conflict of interest.

References

1. Ratliff, P. The new Siemens gas turbine SGT5-8000H for more customer benefit. *VGB Powertech* **2007**, *87*, 128.
2. Ahmadi, G. Evaluation of synchronous execution of full repowering and solar assisting in a 200 MW steam power plant, a case study. *Appl. Therm. Eng.* **2017**, *112*, 111–123. [CrossRef]
3. Zhang, H. Numerical simulation and optimization of directional solidification process of single crystal superalloy casting. *Materials* **2014**, *7*, 1625–1639. [CrossRef]
4. Berkache, A.; Dizene, R. Numerical and experimental investigation of turbine blade film cooling. *Heat Mass Transf.* **2017**, *53*, 3443–3458.
5. Xu, X.Q.; Niu, S.X.; Wang, X.G. Fabrication and casting simulation of composite ceramic cores with silica nanopowders. *Ceram. Int.* **2019**, *45*, 19283–19288. [CrossRef]
6. Niu, S.X.; Xu, X.Q.; Li, X. Microstructure evolution and properties of silica-based ceramic cores reinforced by mullite fibers. *J. Alloys Compd.* **2020**, *829*, 154494. [CrossRef]
7. Kazemi, A.; Faghihi-Sani, M.A.; Nayyeri, M.J.; Mohammadi, M.; Hajfathalian, M. Effect of zircon content on chemical and mechanical behavior of silica-based ceramic cores. *Ceram. Int.* **2014**, *40*, 1093–1098. [CrossRef]
8. Wang, L.Y.; Hon, M.H. The effect of zircon addition on the crystallization of fused silica—A kinetic-study. *J. Ceram. Soc. Jpn.* **1994**, *102*, 517–521. [CrossRef]
9. Wilson, P.J.; Blackburn, S.; Greenwood, R.W.; Prajapti, B.; Smalley, K. The role of zircon particle size distribution, surface area and contamination on the properties of silica-zircon ceramic materials. *J. Eur. Ceram. Soc.* **2011**, *31*, 1849–1855. [CrossRef]
10. Wang, L.L.; Liang, J.; Fang, G.D.; Wan, X.Y.; Xie, J.B. Effects of strain rate and temperature on compressive strength and fragment size of ZrB_2-SiC-graphite composites. *Ceram. Int.* **2014**, *40*, 5255–5261. [CrossRef]
11. Orlovskaya, N.; Stadelmann, R.; Lugovy, M. Mechanical properties of ZrB_2-SiC ceramic composites: Room temperature instantaneous behavior. *Adv. Appl. Ceram.* **2013**, *112*, 9–16. [CrossRef]
12. Yao, Y.J.; Li, C.C.; Wang, L.; Jiang, X.L.; Qiu, T. Mechanical behaviors of alumina ceramics doped with rare-earth oxides. *Rare Met.* **2010**, *29*, 456–459. [CrossRef]
13. Gao, Y.B.; Tang, T.G.; Yi, C.H.; Zhang, W.; Li, D.C.; Xie, W.B.; Ye, N. Study of static and dynamic behavior of TiB_2-B_4C composite. *Mater. Des.* **2016**, *92*, 814–822. [CrossRef]
14. Wu, M.X. Fabrication of TiO_2 components by fields activated sintering technology (FAST). *Ceram. Int.* **2017**, *43*, 8075–8080. [CrossRef]
15. Guo, W. Microstructure and mechanical properties of C/C composite/TC4 joint with inactive AgCu filler metal. *Ceram. Int.* **2015**, *41*, 7021–7027. [CrossRef]
16. Xu, Q.Y.; Xu, Z.L. External Auxiliary Heating and Testing Device for Measuring High Temperature Compression Properties of Materials. ZL 2017 1 0332245.5, 2211201.
17. Pabst, W.; Gregorova, E. Elastic properties of silica polymorphs—A review. *J. Ceram. Silik.* **2013**, *57*, 167–184.
18. Breneman, R.C.; Halloran, J.W. Effect of cristobalite on the strength of sintered fused silica above and below the cristobalite transformation. *J. Am. Ceram. Soc.* **2015**, *98*, 1611–1617. [CrossRef]
19. Weibull, W. A statistical distribution function of wide applicability. *J. Appl. Mech.-T ASME* **1951**, *18*, 293–297.
20. Heimann, R.B. *Classic and Advanced Ceramics: From Fundamentals to Applications*; John Wiley & Sons: Hoboken, NJ, USA, 2010.

21. Mackenzie, J.D. Fusion of quartz and cristobalite. *J. Am. Ceram. Soc.* **1960**, *43*, 615–619. [CrossRef]
22. Scherer, G. A study of quartz melting. *Phys. Chem. Glas.* **1970**, *11*, 53.
23. Wang, L.L.; Fang, G.D.; Liang, J. High temperature tensile damage behavior of ZrB_2-based ultra-high temperature ceramic composites. *Acta Mater. Compos. Sin.* **2015**, *32*, 125–130.
24. Xu, Z.L.; Zhong, J.W.; Su, X.L.; Xu, Q.Y.; Liu, B.C. Mechanical properties and constitutive model of silica-based ceramic core for directional solidification of single crystal superalloy. *Rare Met. Mater. Eng.* **2019**, *48*, 803–806.

Publisher's Note: MDPI stays neutral with regard to jurisdictional claims in published maps and institutional affiliations.

© 2020 by the authors. Licensee MDPI, Basel, Switzerland. This article is an open access article distributed under the terms and conditions of the Creative Commons Attribution (CC BY) license (http://creativecommons.org/licenses/by/4.0/).

MDPI
St. Alban-Anlage 66
4052 Basel
Switzerland
Tel. +41 61 683 77 34
Fax +41 61 302 89 18
www.mdpi.com

Materials Editorial Office
E-mail: materials@mdpi.com
www.mdpi.com/journal/materials

www.ingramcontent.com/pod-product-compliance
Lightning Source LLC
LaVergne TN
LVHW070916170726
843515LV00005B/799